Global TOPS

100 science lessons with 15 simple things

A curriculum that redefines how you teach science!

lab·o·ra·to·ry: any room with tables and chairs.

LABORATORY

stock·room: one box of inexpensive materials.

STOCK-ROOM

TEACHER RESOURCE MANUAL

Conceived and written by
RON MARSON
Illustrated by
PEG MARSON
Lessons on Animals contributed by
DON BALICK

TOPS LEARNING SYSTEMS

342 S Plumas Street
Willows, CA 95988

www.topscience.org

DEDICATION

Global TOPS is the product of 2 cultures plus 20 years of teaching experience. Its origins trace back to Ghana, West Africa, where author Ron Marson first taught math and science as a Peace Corps volunteer. In Africa Ron first learned how to "make do" with limited resources, how to recycle local materials as learning tools.

Returning to America, Ron applied his improvisation skills, developed under third-world conditions of scarcity, to the disposable, throw-away economy of the first-world. He published hundreds of creative activities that reuse simple materials like paper clips, bottles, clothespins and cans to facilitate hands-on learning.

Ron has gathered 100 of his TOPS activites from many published programs, and redesigned them to use a bare minimum of materials. No lab, no utilities, no photocopies required. He calls his program Global TOPS because it works (when translated) anywhere in the world.

Peace Corps Volunteer Ron Marson came to Ghana expecting to teach and to give. The people taught him all about generosity, resourcefulness, creativity and dignity. He received much more than he gave. Global TOPS is Ron's way of returning the love that he received. To the people of Ghana, to poor children everywhere, he dedicates this work.

A MULTI-AGE PROGRAM

Global TOPS works with any grade level *if* you recognize the needs and abilities of your particular age group. LOWER PRIMARY students who are just learning to read should watch you demonstrate activities, and participate as they can. UPPER PRIMARY students can perform most experiments successfully if you supervise them closely and introduce worksheets, as instructed, before students begin independent work. MIDDLE SCHOOL and SECONDARY students require less teacher intervention with increasing age. As students take more responsibility for their own learning, your role as teacher shifts from instructor to consultant. Even COLLEGE students love to use TOPS, especially those training to become science teachers.

ISBN 978-0-941008-91-4

TOPS LEARNING SYSTEMS

Dear Educator,

Consider these inspiring lines from the United States Declaration of Independence:

> We hold these truths to be self evident, that all "children" are created equal; that they are endowed by their Creator with certain unalienable rights; that among these are life, liberty, and the pursuit of "knowledge"...

Yes, we've changed 2 words (in quotes) to emphasize our concerns as educators. But the spirit of this declaration remains unchanged.

While these famous words describe an ideal that people in every nation uphold as self-evident truth, in actual practice, our world is not an equal place. It is painfully clear that God's children do not share equally in Earth's resources. Some have great material abundance and educational opportunity. Others have very little, almost nothing.

Global TOPS is our way of helping to right this terrible injustice. Our goal is to make affordable hands-on science accessible to curious minds everywhere: in industrialized countries and developing countries; in rural schools and inner-city schools; in affluent classrooms and impoverished classrooms; among rich students and poor students. A room with tables and chairs is our universal laboratory. A box with 15 simple things is our global stockroom.

We realize that asking a price for Global TOPS, even a moderate price, selectively discriminates against teachers and schools that can least afford it. It makes students from poor schools "least equal". Yet this same purchase price keeps TOPS in business and Global TOPS in print. What to do?

With your help we believe we can sell Global TOPS on the basis of need rather than profit, and still survive financially. Those of you who pay our normal school prices will not only receive TOPS quality for your money, you'll also enable us to extend the program to poorer schools that can't afford full price. If you are unable to pay the prices we ask, please write and tell us of your special circumstances.

Those of you who teach the very poor can be of special help. A Peace Corps Volunteer serving in a developing country, for example, might draw up a proposal with government officials for translating and printing Global TOPS in-country, in exchange for royalty-free permission to do so. A teacher in the inner city with zero science budget might persuade school administrators to purchase science supplies in return for TOPS providing print materials at cost or free of charge.

We will consider any proposal that demonstrates good faith and personal initiative — a willingness to put Global TOPS into the hands of children who would otherwise never experience science as a process of creative play and self-discovery. We will respond as we are able.

Equal access to education for all children is our collective challenge. Working together we can make a difference.

Peace,

Ron Marson

Ron Marson

A TOPS Model For Effective Science Teaching. . .

If science were a set of explanations and a collection of facts, you could teach it with blackboard and chalk. You could assign students to read chapters and answer the questions that followed. Good students would take notes, read the text, turn in assignments, then give you all this information back again on a final exam. Science is traditionally taught in this manner. Everybody learns the same body of information at the same time. Class togetherness is preserved.

But science is more than this.

Science is also process — a dynamic interaction of rational inquiry and creative play. Scientists probe, poke, handle, observe, question, think up theories, test ideas, jump to conclusions, make mistakes, revise, synthesize, communicate, disagree and discover. Students can only learn these things if they are also free to think and act like scientists, in a classroom that recognizes and honors individual differences.

Science is *both* a traditional body of knowledge *and* an individualized process of creative inquiry. Science as process cannot ignore tradition. We stand on the shoulders of those who have gone before. If each generation reinvents the wheel, there is no time to discover the stars. Nor can traditional science continue to evolve and redefine itself without process. Science without this cutting edge of new discovery is a static, dead thing.

Here is a teaching model that combines the best of both elements into one integrated whole. It is only a model. Like any scientific theory, it must give way over time to new and better ideas. We challenge you to incorporate this TOPS model into your own teaching practice. Change it and make it better so it works for you.

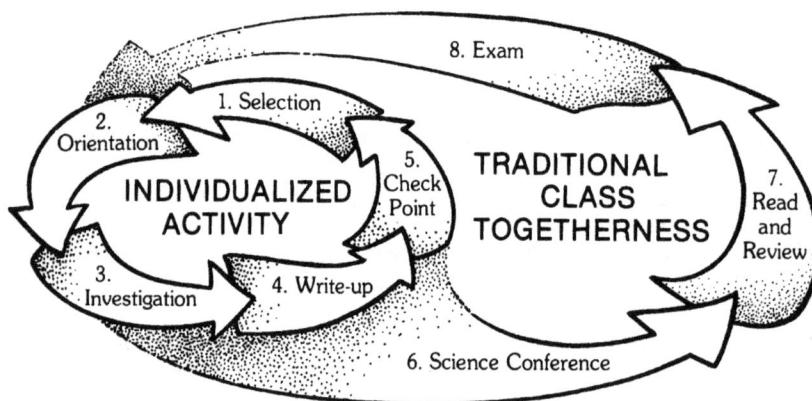

1. SELECTION

Decide which of the 15 possible topics of study, A through O, you wish to study next. The alphabetical ordering, listed in the table of contents, is but one of many possible sequences. Your school's teaching syllabus might dictate a different arrangement. Better yet, let students decide what they want to study next. Consult "Sequencing" in the introduction to each topic, to determine what, if any, prerequisites must be completed first.

Once you have selected a topic, begin the individualized activity cycle. Students simply turn to the appropriate page in their *Student Reference Books* and complete each activity in sequence on their own papers. Teachers not wishing to use student reference books can also photocopy each activity in class quantities. The *Teacher Resource Manual* contains full-sized reproducible activity sheets for this purpose.

Those who finish a topic early should be encouraged to do original investigations that go beyond assigned activities. "Extension" sections in many of the teaching notes, plus "Further Study" listings in the introduction to each topic, provide creative jumping-off points to open-ended exploration.

2. ORIENTATION

Clear directions and detailed illustrations enable students to interpret worksheet instructions on their own. This allows students to start each activity when they are ready to start it, not when the teacher has time to explain it.

Identify poor readers in your class. When they ask, "What does this mean?" they may be asking in reality, "Will you please read these instructions aloud?" You can help poor readers by pairing them with good readers, by encouraging students to help each other.

Some activities deal with concepts that are especially difficult (often math-related). They require extra explanations that only a teacher can provide. Check the teaching notes under "Introduction" to see what teacher input, if any, is required. Just before your most advanced students start an activity that requires a special introduction, call a temporary halt to individual activity so you can teach your whole class at once. Students can then apply what you've taught them individually, as each one completes the activity.

3. INVESTIGATION

Worksheets structure independent class activity. Students do experiments as directed using simple materials stored on shelves and in boxes. The grids, rulers, or other graphics required for some experiments should be removed from the consumable *Student Cutout Booklets*. These may be conveniently stapled to the back of each student's science notebook or assignment folder.

Students should think and act on their own and help each other. You'll need to answer questions and provide assistance, to be sure, but do so only *after* students have first tried to solve their problems independently. Make a conscious effort to stay out of the center of attention.

As you teach each lesson from one year to the next, new and better ways of doing things will gradually become apparent. Be sure to record your own better ideas, as you think of them, on the same page as the rest of the teaching notes. There is ample white space on each page for this purpose.

4. WRITE-UP

Activity sheets ask students to explain the how and why of things. Students respond on their own papers, never in the *Student Reference Book*. Reference books are for reading *only*. They will last many years if you take proper care of them. They are a welcome low-cost alternative to expensive photocopying.

Keep all write-ups on file in class. Notebooks, copybooks or file folders all serve as suitable assignment organizers. Students will feel pride and accomplishment as they see their file folders grow heavy, or their notebooks fill up, with completed assignments. Having all papers in one place also facilitates easy reference and convenient review.

Ask students to make an assignment record and staple it to the front of each notebook or file folder. They should list all 100 lessons in alphabetical order (from A-1 to O-5), on a single sheet of lined paper in 3 or 4 columns. As students complete each assignment, you simply initial your approval next to the corresponding assignment number. This record tells you at a glance if each student is working on schedule.

5. CHECK POINT

As lessons are completed, students should bring their personal write-ups to you for evaluation. (Make a class rule that the next activity should not be started until you have initialed your approval of the last one on their assignment records.) The student and teacher then evaluate these write-ups together on a pass/no-pass basis. You'll find an answer key for each lesson listed under "Check Point" in the teaching notes.

Because students are present when you evaluate, feedback is immediate and effective. A few seconds of direct student-teacher interaction is surely more effective than 5 minutes worth of margin notes that students may or may not read. Remember, you don't have to point out every error. Zero in on particulars.

If answers are wrong or write-ups are incomplete, direct students to make specific improvements. They should see you again for another check point when these improvements have been made.

6. SCIENCE CONFERENCE

Set a limit to the number of days you'll dedicate to individualized process science (usually 1 day per lesson). Announce the deadline for turning in all assignments well in advance. This gives your class sufficient warning to make an extra effort to get everything turned in on time. Not everyone will meet your deadline, of course. Too bad. Individualized activity has ended.

Science Conference is a time for students to come together, to discuss experimental results, to debate and draw conclusions. Those who did original investigations or made unusual discoveries share this information with their peers, just like scientists at a real conference.

This is also a good time to consider the technological and social implications of the topic you are studying. Invite speakers from your community to come in and talk to your class. Read and discuss newspaper articles of interest. Show relevant films. Consider possible resources available in your school and wider community, then bring them together.

7. READ AND REVIEW

Does your school have an adopted science textbook? Do parts of your science syllabus still need to be covered? Now is the time to integrate other traditional science resources into the overall program.

Your students already share a common background of hands-on lab work. They have discussed relevant social issues. With this shared base of experience, they can now read the text with greater understanding, think and problem solve more successfully, and communicate more effectively.

You might spend just a day on this step or an entire week. It all depends on your particular teaching style, the concepts you wish to cover and the resources you have available. The introduction to each topic contains a page of suggested evaluation questions. Keep these questions (plus others of your own choosing) firmly in mind as you review key concepts in preparation for the exam that follows.

8. EXAM

Use any combination of test questions, plus questions of your own, to determine how well students have mastered the concepts you've been teaching. Students who finish your exam early might begin work on the first activity in the next new topic.

Base your overall grade on each student's overall performance. We recommend that you give these 3 evaluation components equal weight: (1) Effort: How many assignments received your check-point approval? (2) Attitude: Did the student participate actively or simply waste time and copy the results of someone else? (3) Achievement: How well did each student demonstrate mastery on the test just completed?

Now that your class has completed a major TOPS learning cycle, it's time to start again with a brand new topic. This is an opportunity to start fresh, to learn from past mistakes, to do better than before. Because each topic is relatively short, your teaching tempo remains brisk and lively; no chance for students to procrastinate or get bored. The frequent change of pace insures that your students will work hard, love what they learn, and grow in scientific literacy.

GATHERING MATERIALS

This page lists *everything* you'll need to complete all 100 TOPS activities. To understand which of these materials are specifically used in any given topic A through O, please consult the appropriate sublist in the introduction to each topic.

Materials to Purchase

Gather everything in a single afternoon, or add materials as you teach. If you follow an A through O alphabetical sequence, then all items are listed in order of first use. Thread is needed first, for example; dry-cells are not required until activities H; nails are first used in activities M.

Quantities are sufficient to support 30 students working in lab groups of 2. In 1988 US dollars, expect to spend roughly $140.00 if you purchase everything on this list in the recommended quantities. Spend about $75 per year thereafter to resupply consumables. Dry cells and bulbs *alone* account for over half of this total expense. Cutting back on these 2 basics will dramatically lower your overall costs.

1 spool — **Thread**.

15 pair — **Scissors**.

11 boxes — **Paper clips** of uniform size and weight. Use one brand only, medium size, about this large. At 100 paper clips per box, you need 1,100 total.

5 rolls — **Masking tape**: 3/4 in x 50 yds or longer (1.9 cm x 46 m).

120 — Medium-sized **rubber bands**.

8 rolls — **Clear tape**: 1/2 in x 12 1/2 yds or longer (1.3 cm x 11 m).

1 box — Steel **straight pins**. Aluminum pins must not be substituted.

90 — Wooden spring-action **clothespins**.

2 rolls — **Aluminum foil**: 12 in x 25 yds (30 cm x 23 m)

60 — Size-D **dry cells** (1.5 volts). Sold most economically in boxes of 36. Paying extra for higher quality is worth your money.

40 — Flashlight **bulbs**. Bulbs with collars are better than screw-in kinds. Use a size that is designed for 2 dry cells (3 volts).

1 box — Fine grade **steel wool**. Do not substitute soaped pads.

70 — Small ceramic **magnets**, with or without a center hole. These are generally available in bulk from science supply outlets at reasonable cost. In the USA, try Radio Shack.

1 roll — Plastic-insulated **copper wire**: approx. 24 guage, 200 ft or longer (61m). Diameter is not critical, but the wire should be thin enough to easily bend back and forth. Avoid bell wire with a baked-on enamel finish.

30 — Medium-sized **nails** about 2 1/2 in (6.5 cm) long. Size is not critical.

Materials to Recycle

In addition to the 15 basics listed above, review this inventory to determine (a) what additional materials (if any) you need to purchase; (b) what you already have on hand; (c) what recyclables your students can bring from home. Notice that some items are optional, while others are used in only a single experiment. Materials are again listed in order of first use.

Quantities are sufficient to support 30 students working in lab groups of 2.

1 — Wall **clock** with second-hand sweep. Your students can substitute wristwatches if they have them.

30 — Small **coins** (pennies) of uniform size.

15 — Hand **calculators** (optional). Mental long division and multiplication is probably more beneficial.

30 — Medium-sized tin **cans**.

sm pkgs — Local **seeds**. Use pinto beans, popcorn, lentils and long-grained white rice, or equivalent. See "Preparation" in teaching notes D-3 and O-1.

15 — **Bottle caps**. Different sizes are OK.

15 — **Pen caps**. Different sizes are OK.

1 handful — **Scratch paper**. Some sheets should be the same size.

1 — A **water** source plus buckets or other large containers for easy distribution.

30 — **Small glass jars with lids**, 1 pint (500 ml) or smaller. At least 15 lids must fit tightly.

15 — **Large glass jars**, 1 quart (1000 ml) or larger.

1pkg — Refined **sugar**. (Used in 1 activity.)

1pkg — Refined **salt**. Seal tightly against moisture for easy pouring. (Used in 1 activity.)

various — Local **insects and animals** to observe.

1 ball — **String**. (Optional for 1 activity.)

various — A **coloring system**: Paints, crayons, or colored pencils. (Used in 1 activity.)

1 handful — **Newspaper**.

1 bucket — **Soil**. Use local variety or purchase a small bag of potting soil.

1 sq yd — **Plastic wrap**. Use plastic bags or a roll of plastic wrap.

School Supplies

Every student is individually responsible to turn in a separate report for each completed activity. As such, each one must have the following basic supplies.

1 — A **system to organize and store assignments**. This might be a spiral-bound or loose-leaf notebook, a copy book, or notebook paper stored in a file folder.

1 — A **pencil**.

1 — An **eraser**.

CONTENTS

LONG-RANGE OBJECTIVES

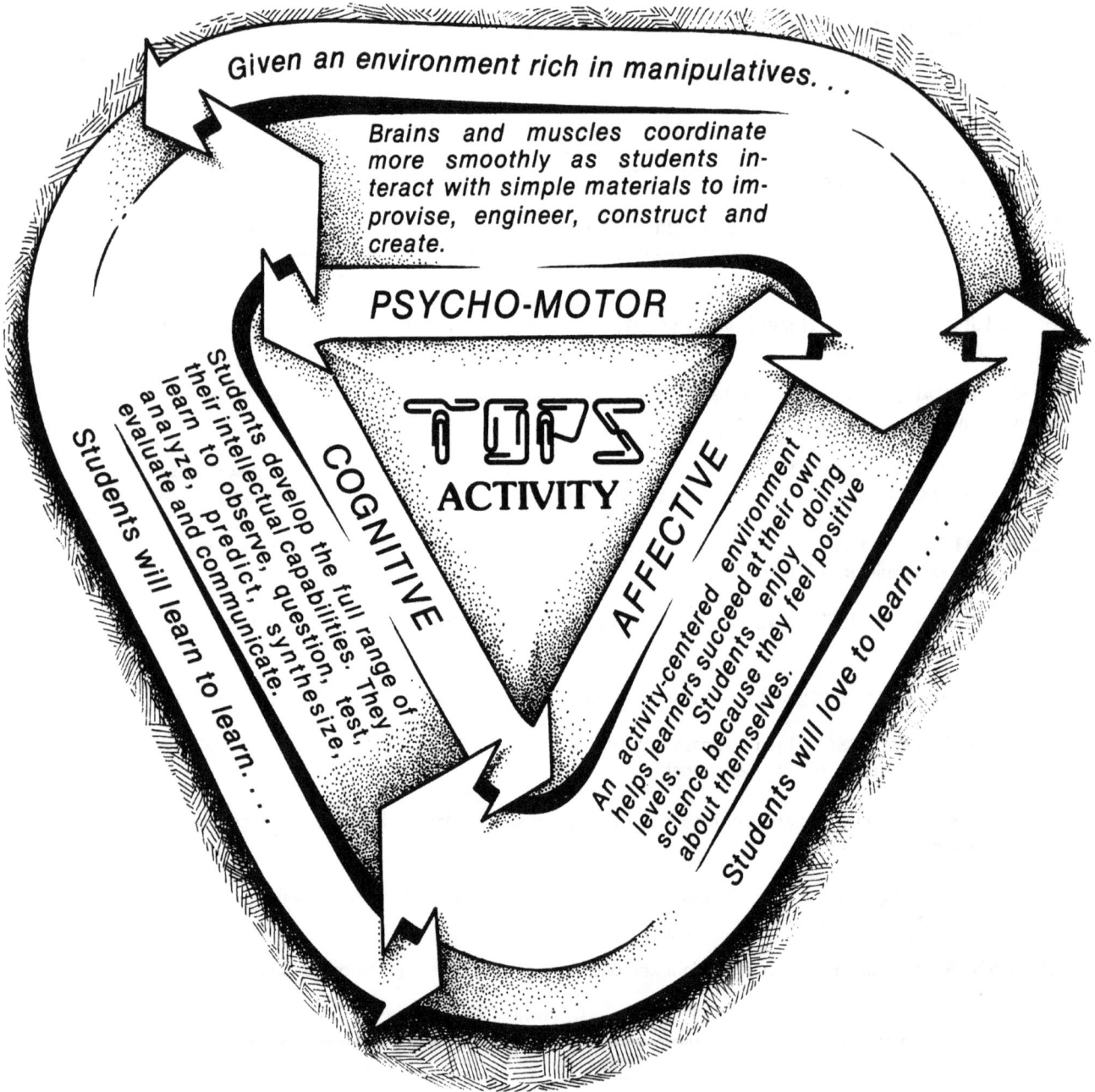

Given an environment rich in manipulatives. . .

Brains and muscles coordinate more smoothly as students interact with simple materials to improvise, engineer, construct and create.

PSYCHO-MOTOR

TOPS ACTIVITY

COGNITIVE

AFFECTIVE

Students develop the full range of their intellectual capabilities. They learn to observe, question, test, analyze, predict, synthesize, evaluate and communicate.

An activity-centered environment helps learners succeed at their own levels. Students enjoy doing science because they feel positive about themselves.

Students will learn to learn. . . .

Students will love to learn. . . .

A. INVESTIGATING PENDULUMS

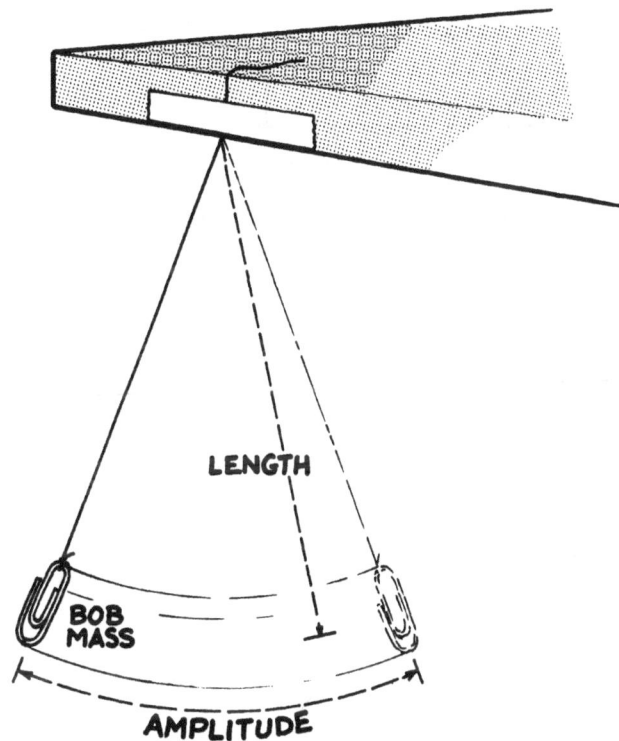

Pendulums are easy to build, as easy as tying thread to a paper clip. They swing back and forth with great regularity, measuring time in clockwork fashion.

Galileo is credited with first discovering that neither the mass of a pendulum's bob, nor its amplitude affect how fast a pendulum swings. Length is the only variable that matters. In these experiments, your students will follow Galileo's good example. They will learn all about pendulums, not by reading what other great scientists have done, but by doing great science themselves.

Finally your students will link pendulums together, then observe how they transfer kinetic energy from one pendulum bob to the other. This behavior is not at all intuitive. Pendulums surprise us by *not* doing what we think they should, and *not* doing it consistently. They teach us to observe the way things are, not the way we think they should be.

━━━━━ EVALUATION ━━━━━

Each question evaluates a single activity from INVESTIGATING PENDULUMS as numbered. Use any combination to frame a formal exam or an informal review: Copy these questions on your blackboard, construct your own ditto master, or photocopy the questions while masking out the rest of the page. Evaluate in ways that suit your own teaching style, enabling your students to learn and enjoy science.

Questions

A-1

a. Which pendulum swings fastest?
b. Which pendulums swing together?
c. Which pendulum swings slowest?

A-2
Your grandfather clock runs too slow.

How would you adjust the clock's pendulum to make it keep better time?

A-3

A heavy person swings on a swing in the park. Then a light person uses the *same* swing.

Do both people swing at the same frequency? Explain.

A-4

Suppose you are swinging from an overhead bar.

a. If you took bigger swings, could you slow your frequency?
b. How could you best increase your frequency?

A-5

Suppose you hang three strips of wood from a table like this.

Which one swings the fastest? Why?

A-6
Two pendulums with equal lengths are taped to a long straw like this.

If you start B swinging toward A, predict what happens.

A-7

Two pendulums with different lengths are taped to a long straw like this.

If you start B swinging toward A, predict what happens.

Answers

A-1
a. E   b. A and D   c. B

A-2
Shorten the pendulum's length. This makes the grandfather clock tick faster.

SLOWER ↓   ↑ FASTER

(Many pendulum clocks have an adjustable screw below the bob that allows you to make slight changes in the overall effective length.)

A-3
Yes, both swing at the same frequency. The swing moves like a pendulum. Its frequency changes only with length, not bob weight (the weight of the person using the swing).

A-4
a. No. The frequency of your body, swinging like a pendulum, does not vary significantly with amplitude.
b. Draw your legs up near your stomach to shorten your overall length.

A-5
Pendulum C swings the fastest. Its length, as measured from the pivot to the *center* of the wooden bob, is the shortest.

A-6
Pendulum B pushes against the straw and begins to transfer its energy of motion through the straw to pendulum A. A swings with greater and greater amplitude until it reaches a maximum, while B slows to a standstill. Then the process reverses and A transfers its energy back to B.

A-7
Pendulum B pushes against the straw and begins to transfer its energy of motion through the straw to pendulum A. But A has a lower frequency than B and soon gets out of phase. So A transfers the energy back again without reaching its maximum amplitude. In effect, B swings continuously while A starts and stops.

SEQUENCING

INVESTIGATING PENDULUMS is a good place to begin. The activities are easy to do and lead logically into B. Material requirements are extremely modest and easy to organize.

Related Activities: **A**---B

MATERIALS

Here is everything your students will use for the next 7 activities on INVESTIGATING PENDULUMS. Materials printed in normal type are part of the core 15-things-in-a-box inventory that support all 100 activities. Materials printed in *italics* are additional local materials that you provide or ask your students to bring from home. Pencil and paper are already assumed and therefore unlisted. Each item is numbered with the activity where it is first used.

(A-1) Thread. Don't substitute string.

(A-1) Scissors.

(A-1) Paper clips of uniform size and weight.

(A-1) Masking tape. Don't substitute clear tape. It's too hard to see and clings tenaciously to desk surfaces. This makes clean-up difficult.

(A-2) A *wall clock* with second-hand sweep. Your students can also use wristwatches if they have them.

(A-6) Rubber bands.

(A-6) *Small coins* of equal size.

FURTHER STUDY

Use problems like these plus "extension" ideas in INVESTIGATING PENDULUMS to lead your students beyond worksheet activity into original research and investigation. Each discovery leads to more questions, deeper questions, better questions than these. Answering them is what good science is all about.

Read about the history of time in an encyclopedia. Sketch examples of different types of early clocks, showing how they worked. Did our ancestors use other things besides pendulums to keep time? Build a clock of your own as a science project.

In politics you hear that the pendulum has "swung in the other direction" or that "events have come full circle". Do you think that history, like the pendulum, repeats itself? Write an argument supporting your position.

Do you like to ready scary stories? Edgar Allen Poe, an American writer of the early 19th Century, was a master of the macabre. Go to the library and read one of his more famous short stories "The Pit and the Pendulum". Write a book report.

NAME: CLASS:

OBSERVING PENDULUMS

1 Cut a piece of thread as long as your arm.

Tie a paper clip to one end.

TRIM

2 Tape it to the bottom edge of your table like this.

LEAVE END FREE

3 Pull the paper clip to one side, then let it go.

Describe how the paper clip moves.

4 Find a way to make your pendulum swing faster; then slower. Tell how you did this.

?? *What should I change?*

5 Hang another pendulum next to your first.

6 Make both pendulums "march together" for at least 10 cycles back and forth.

Release together.

☐ TEACHER CHECK

Tell how you did this.

7 Predict how these pendulums will swing. Give reasons for each answer.

A

B

C

D

Fastest? Slowest? The same?

Objective

To observe a pendulum and describe its motion. To understand, in qualitative terms, how the frequency of a pendulum changes with length.

Introduction

The activity sheets in this book are reproducible, should you elect to photocopy. There is, however, a more economical hassle-free approach. Give your class a set of *Student Reference Books* instead. Remind them that these books are for reference only — to be shared by everybody — now and in years to come. No writing or cutting allowed! Students respond on their own assignment papers and cut out the required graphics from their own consumable *Student Cutouts Booklets*.

Lesson Notes

1. Young children may experience difficulty when attempting to knot thin wispy thread. Remind them to use 2 loops, not 1. Explain that big loops are easier to tie than small loops. Identify students who are good at tying knots, and ask them to help those who have difficulty.

2. Some students may tape the pendulum to the *top* surface of the table rather than the *edge*. From a practical point of view this is OK. Experimental results are not the least bit affected. However, there is a technical reason for using the edge. This gives the pendulum a single stationary pivot. Notice how the pivot jumps back and forth between two different points when the thread is taped to the top table surface.

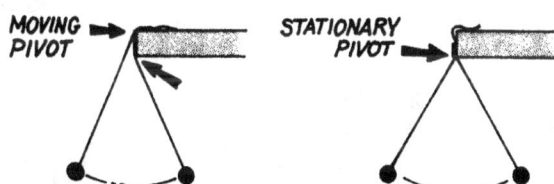

4. To lengthen or shorten the pendulum, simply pull the thread *through* the tape. Students don't need to reposition the tape, although some may do so.

Allow students to discover the relationship between length and frequency in their own good time. Don't rob them of the joy of discovery by telling them the "correct" answer before they've had a chance to do their own experimenting.

5. Bring the second piece of tape even with the first, flush with the bottom edge of the table. This insures that both pendulums will hang from the same height.

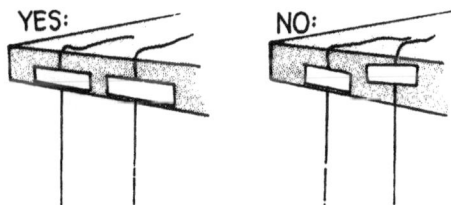

6. An easy way to start both paper clips simultaneously is to release them from a ruler, a book, or the back of your hand.

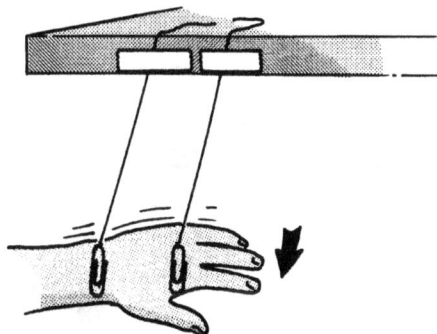

It is not possible, of course, to make pendulums (or anything else) *exactly* the same length. The best students can do is stay within a specified tolerance. In this case the lengths must be adjusted so nearly equal that the pendulums still remain "in step" after ten cycles. This demands very subtle length adjustments. As your students struggle to do this, they'll appreciate first-hand the sensitive interdependence between pendulum length and frequency.

Those who can't meet this tolerance might be directed to use longer pendulums. (Long pendulums are less sensitive to subtle changes in length than shorter pendulums.) Since *you* administer the teacher check, *you* decide when each student has done his or her best and earned the privilege to move on to step 7.

7. Make sure your class understands what it means to "predict". (Make an educated guess based on previous experience.) If students are unsure of their predictions, or have guessed wrong, ask them to actually set up the pendulums as illustrated and check their answers.

Extension

Hold a class contest to see who constructed the best pendulum pair. Ask everyone to release their pendulums at the same time. You, or a group of student judges then identify those pairs that remain in step over the longest time. Match these finalists against each other to determine a class champion.

Check Point

3. The paper clip swings back and forth through an arc. It moves fastest through the bottom of its arc, slows to a complete stop at the top, then accelerates back down in the opposite direction. (The paper clip also describes an ellipse when swung in a circle.)

4. To make a pendulum swing faster, decrease its length; to slow a pendulum, increase its length.

6. Adjust each pendulum until they have the same length.

7. Pendulum A swings fastest. Pendulum C swings slowest. Pendulums B and D swing with the same frequency.

CLOCK PENDULUMS

1 Stand up when you think *exactly* one minute has passed. Your teacher will keep time to see who comes closest.

Put all watches out of sight...

... get ready, get set, GO!

2 Now make a pendulum clock that ticks exactly 60 seconds in a minute.

a. Tie an arm-length thread to a paper clip.

b. Lightly fold a small piece of masking tape over the middle of the thread.

PULL THROUGH

c. Slide the thread through the tape until you find a length where your pendulum swings exactly *one cycle each* second.

1 second , 2 seconds, 3 seconds ...... 60 seconds

3 Repeat the class experiment in step 1. This time use your pendulum clock to keep time.

Hide your watches! Ready, set, GO!

4 Did your pendulum clock help you guess one minute more accurately? Explain.

Objective

To make a pendulum that ticks out time like a clock — 60 cycles per minute.

Introduction

Begin by leading your whole class in step 1. As the teacher, you play an essential time-keeping role. Be prepared to repeat this process in step 3.

Lesson Notes

1. Students attempt to guess how long one minute lasts while you keep time. To put everyone on an equal footing (except you, the timekeeper), be sure that all wristwatches are placed in pockets and all wall clocks are out of view. Begin with everyone seated. Tell students to stand when they think exactly 60 seconds have passed. Note who stood closest to the mark, but don't declare the winner until everyone is standing.

2. Masking tape makes an ideal pendulum pivot. Simply hold it in your hand to make the pendulum swing. The tape won't move relative to the thread, until you intentionally decide to pull it back or forward to create a new pendulum length. Through trial-and-error, it is possible to find just the right length adjustment, so the pendulum marks time with impressive accuracy.

Notice that the pendulum ticks out seconds in cycles (once back and forth), not swings. To investigate clocks that mark time in swings, see the "Extension" below.

3. Students guess the length of 1 minute again, this time aided by their pendulum clocks. If these clocks are accurate, your entire class should rise, in choir-like unison, near the 60 second mark.

Extension

Develop a slower clock pendulum that makes 1 *swing* every second.

a. Measure how many times the shorter *cycle*-per-second pendulum fits into this longer *swing*-per-second pendulum? (Four times. This relationship will be explored in greater detail in activity B-2.)

b. How long would it take your swing-per-second pendulum to complete 1 million swings?

1,000,000 s x 1 min/60 s x 1 hr/60 min x 1 day/24 hr
=11.57 days

Check Point

4. Yes. I estimated much closer to the 60 second mark when using the pendulum clock. So did the rest of the class because we all stood up nearly together at the 60 second mark.

NAME: _____ CLASS: _____

HEAVIER OR LONGER?

1 Cut some thread as long as this paper. Tape it to the edge of your desk so it forms a loop.

2 Hang a paper clip from this loop to make a pendulum.

3 Count how many cycles your pendulum makes in one minute.

1, 2, 3...

Frequency?

Double check your answer.

4 Now make your pendulum *heavier*. Hang 10 paper clips from the loop.

10 CLIPS THROUGH THE LOOP!

Frequency?

Double check your answer.

5 Now make your pendulum *longer*. Arrange the 10 clips to form a chain.

ONE LONG CHAIN!

Frequency?

Double check your answer.

6 To slow a pendulum (make it swing with less frequency) should you make it *heavier* or *longer*? Use numbers to support your answer.

Objective

To determine if changes in bob weight affect the frequency of a pendulum.

Introduction

Write these pendulum terms on your blackboard.

CYCLE   FREQUENCY

A **cycle** is two swings, once back and once forth.

The **frequency** of a pendulum is the number of cycles it makes in a minute.

Illustrate what these words mean by making a paper clip pendulum, counting cycles over a one minute interval, and thus finding its frequency.

Lesson Notes

1-2. Because the thread is doubled, it forms a short pendulum, reaching only half the length of this page. This creates relatively high frequencies in steps 3 and 4 that slow dramatically when the pendulum is lengthened in step 5. This drop in frequency would not be as obvious if students started with a longer pendulum.

3. Scientists express how fast pendulums swing in two different ways: by measuring the *period*, the number of seconds required to complete 1 cycle; or by finding the *frequency*, the number of cycles counted during a specific time interval. Because frequencies are typically measured over longer time intervals (1 minute rather than seconds), they are expressed as large whole numbers. Periods by contrast are usually expressed as decimals. Suppose, for example, a pendulum requires .8 seconds to complete each cycle. Then its frequency is 75 cycles/minute, whereas its period is .8 seconds/cycle.

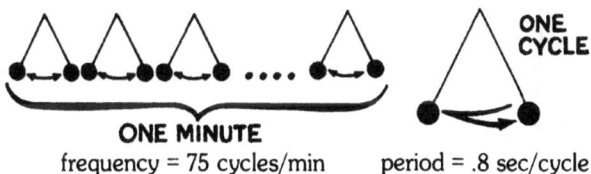

frequency = 75 cycles/min   period = .8 sec/cycle

If you were a novice science student, which number would you rather deal with? Frequency is our choice, too. The numbers are easier to think about, to measure, and to express.

3-5. To accurately measure frequency, it is necessary to look simultaneously at the pendulum (to count cycles) *and* the clock (to measuring time). Pairs of students might conveniently divide this responsibility — one keeps track of time while the other counts. To minimize errors, ask students to double-check each answer. If they get the same value twice, chances are good they have made an accurate determination.

Numbers without units are meaningless. They could represent coconuts, oranges, chickens, anything at all. Insist that your students specify *what* they are measuring in each answer: they should write "cycles/minute," or "c/m" for short. The frequency values in steps 3 and 4 should agree to within a few cycles per second. Any difference between them is simply a matter of experimental error. Those who are not familiar with pendulums, or haven't observed them closely, will find it hard to believe that light and heavy pendulums swing with the same frequency. Only by changing the length in step 5 (rearranging the paper clips to form a long chain) will the frequency drop significantly. Length, not weight, is the only variable that matters.

Check Point

Student values may differ from these, depending on the final length of the pendulums they measure. In general, steps 3 and 4 must have nearly equal values, and the value in step 5 must be significantly less.

3. Frequency = about 80 c/m.

4. Frequency = about 80 c/m.

5. Frequency = about 54 c/m.

6. To reduce frequency, make the pendulum longer. When 10 clips were hung in a long chain the frequency dropped from 80 c/m to 54 c/m. It was increased length, not weight, that made this difference. When the 10 clips were added in a bunch, the frequency remained unchanged at 80 c/m, despite the added weight.

NAME: _____ CLASS: _____

LITTLE SWINGS / BIG SWINGS

1 Make a paper clip pendulum so it hangs about half way to the floor.

ABOUT HALF-WAY UP.

2 Pull out the arm of the paper clip to make a hook.

3 Add about 10 clips to the hook to make a heavy pendulum bob.

THE WEIGHT IS CALLED THE BOB.

4 *Amplitude* is the amount of swing. Find the frequency (in cycles per minute) for these 3 amplitudes.

36, 37, 38...

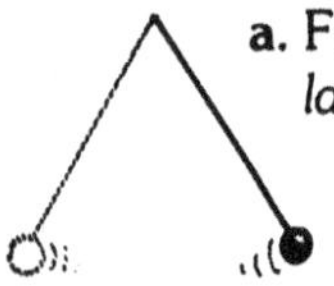

a. Frequency for *large* amplitude.

b. Frequency for *medium* amplitude.

c. Frequency for *very small* amplitude.

Double check any answer you are not sure about.

5 When you change the amplitude, does the frequency change by much? Include frequency values in your answer.

6 Tell how each variable changes the frequency of a pendulum (if at all).

a. **LENGTH**

b. **BOB WEIGHT**

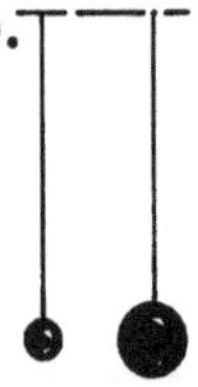

c. **AMPLITUDE**

Objective

To determine if changes in amplitude affect the frequency of a pendulum.

Introduction

None required.

Lesson Notes

4-5. Notice that students are asked to double-check any answer they are not sure about. By doing this, they are less likely to obtain frequency values that differ significantly, then erroneously attribute these differences to changes in the amplitude variable.

This does not mean that all three frequency values will be identical. Very large amplitudes do, in fact, slow a pendulum perhaps a cycle or two when measured over a one-minute time interval. (Medium to small amplitude changes cause no fluctuation.) More important, errors in measurement can be reduced by exercising care, but never eliminated completely. No matter how carefully you measure, you always reach a limit of certainty, beyond which your instruments and ability cannot go.

Be sure to praise students when they report unavoidable measuring errors. Be suspicious about data that looks too perfect. Otherwise your students will learn to please you with what "ought to be" rather than reporting "what is."

6. Some may be unfamiliar with the term "variable." A variable is any physical property that changes. It may or may not affect the outcome of an experiment. Short pendulums swing at higher frequencies than long pendulums, but amplitude, bob weight, and the population of Mexico City are three variables that do not matter.

Students are typically slow to accept the result that amplitude and bob weight do not affect frequency outcomes. It seems the pendulums aren't behaving as they should. (They always do, of course.) You can offer convincing proof by doing this simple demonstration: Suspend two paper clip pendulums from thread loops so they hang in plain view of your whole class. Lengthen or shorten as necessary so they swing with the same frequency; that is, march together in perfect step when you release them together.

Now add more paper clips to one of the loops. Will both pendulums still march in step? Absolutely! The multi-clip pendulum has greater bob weight *and* (because it comes to rest more slowly) greater amplitude. But the single-clip pendulum still beats out the same time, marching right beside in mini-steps of low amplitude. Neither the bob weight nor amplitude variables affect frequency.

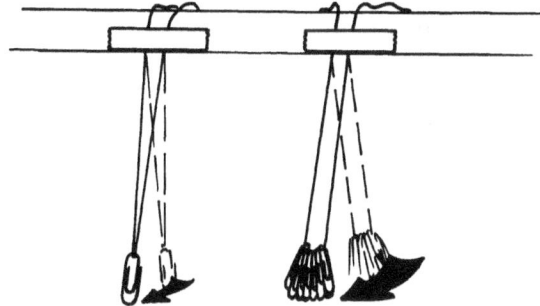

Check Point

Student values may differ from these, depending on the exact length of the pendulums they measure. Nevertheless, all three answers should be nearly equal.

4a. Frequency for LARGE Amplitude = 48 c/m.
 b. Frequency for MEDIUM Amplitude = 49 c/m
 c. Frequency for SMALL Amplitude = 49 c/m

5. The amplitude has little effect on pendulum frequency. Pendulums moving through large, medium and small amplitudes all had frequency values that remained relatively unchanged at 48 to 49 cycles per minute.

6a. Shorten a pendulum to increase frequency. Lengthen a pendulum to decrease frequency.
 b. Bob weight has no effect on frequency.
 c. Amplitude has little effect on frequency. (Large amplitudes may reduce the frequency slightly.)

HOW LONG IS A PENDULUM?

1 *Predict*: Will 5 clips swing faster than 1 if you line up the *tops* of the pendulum bobs?

a. Tell why you think so.

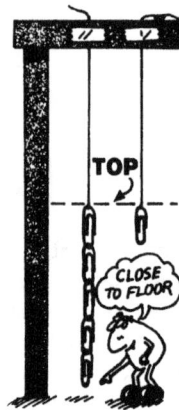

b. Test your prediction. Then write your results.

TOP

CLOSE TO FLOOR

2 Predict: How will the pendulums swing if you line up the *bottoms* of the bobs.

a. Tell why you think so.

b. Test your prediction. Then write your results.

MAKE THIS ONE LONGER

BOTTOM

3 Find a length where both pendulums swing together.

a. Draw how they look.

b. Where do the bobs line up?

4 Hang a chain with 3 clips next to your other 2 pendulums. Make all 3 swing together.

a. Draw how they look.

b. Where do the bobs line up?

5 Pendulums with the same length do swing together. But you must measure from the pivot to. . .

PIVOT

6 Pendulums A, B and C are all taped to the edge of a table. Which 2 swing together? Why?

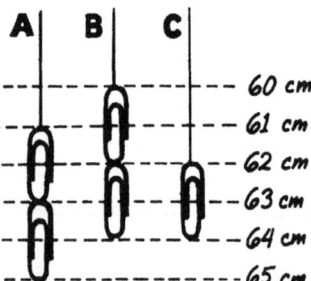

A **B** **C**

- - - - - - 60 cm
- - - - - - 61 cm
- - - - - - 62 cm
- - - - - - 63 cm
- - - - - - 64 cm
- - - - - - 65 cm

Objective

To understand why the length of a pendulum must be measured from the pivot to the center of the bob.

Introduction

None required.

Lesson Notes

1. Your students should easily recognize that the 1-clip pendulum swings faster than the 5-clip pendulum. That's because it's shorter, of course. But watch out for those few who still claim it swings faster because it's lighter! Old mindsets are hard to overcome.

Notice that the 5-clip pendulum almost touches the floor. This is important. By keeping the thread as long as possible relative to the bob, you can treat this pendulum as ideal — that is, as a weight concentrated at a single point source swinging from a weightless line. Non-ideal pendulums follow a more complicated physics that don't obey the idealized generalizations in steps 5 and 6.

Throughout this activity, students compare pendulum lengths by observing how the paper clip bobs hang side by side. This only works if both pivots are placed at equal heights; that is, if the pendulums are taped side by side to the bottom edge of the table.

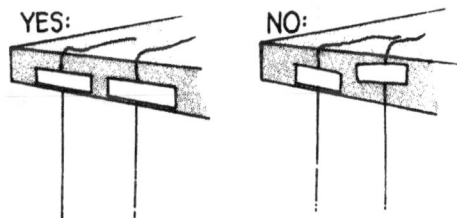

As usual, adjust the pendulum lengths by pulling the thread through the tape. Don't reposition the tape.

2. It seems these two pendulums should swing together because they have the same length. Wrong! They don't have the same length, not if you measure from the pivot to the bob's center of mass. Steps 3 through 6 resolve this apparent contradiction.

3-4. If the pendulums swing together yet don't quite line up along the middle of their paper clip bobs, two things may be wrong. First check to see that the pivots are positioned at equal heights along the bottom table edge. Next make sure the 5-clip pendulum extends almost to the floor (make it as long, and therefore as ideal, as possible).

Check Point

1a. The 5-clip pendulum should swing slowers because it is longer.
 b. My prediction was correct.

2a. The pendulums should swing together because they have the same length. (Even though this prediction is wrong, let it stand.)
 b. My prediction was wrong.

3.

The 2 bobs line up in the middle.

4.

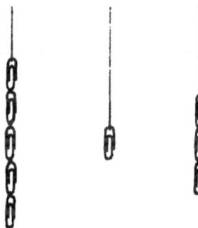

The 3 bobs line up in the middle.

5. . . . the center of the bob.

6. A and C swing together because they have the same length as measured from the pivot to the bob center.

ENERGY TRANSFER

1 Tape 2 coins to pieces of thread. Hang them from your table like this:

Twice as long as this paper.

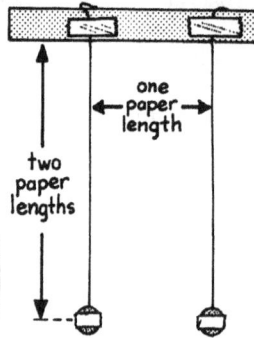

one paper length

two paper lengths

2 Loop each pendulum over a rubber band so it rests about halfway up.

Loop each side.

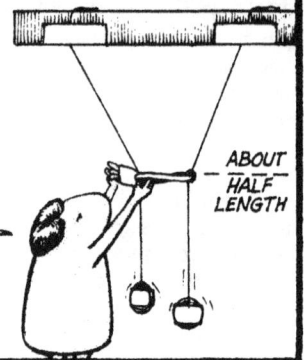

ABOUT HALF LENGTH

3 While 1 coin hangs still, pull back the other and let it swing.

Do the swinging coins transfer energy? Explain how you know.

4 What happens as you move the rubber band up or down?

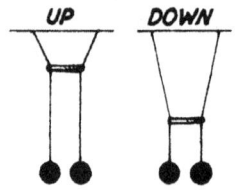

UP   DOWN

5 *Predict*: Can *one* pendulum transfer *all* its energy to the other if they have different lengths? Explain your reasoning.

REMEMBER, Each pendulum swings at a different frequency.

6 Test your prediction. How do the pendulums interact?

Objective

To observe energy transfers in coupled pendulums of equal and unequal length. To appreciate that a pendulum optimally accepts energy when it is delivered in phase with its natural frequency.

Introduction

Remember as a small child when you couldn't make a swing move unless you were pushed? Taking a ride always depended on the generosity of someone else. Then one day you figured out how to push yourself. The secret was to *wait* until you were at the height of your backswing, then take a sharp dip. You finally understood, at least on an intuitive level, that a swing only accepts energy when delivered in phase with the its own natural frequency.

Pendulums respond just like your childhood swing. (Make a short and a long pendulum to demonstrate.) You must push the long pendulum slowly, in time with its own natural frequency, to give it energy. You must push the short pendulum rapidly, in time with its higher natural frequency, to give it energy. If energy pulses are not delivered in phase with each pendulum's own natural frequency, it won't be able to accept all the energy being delivered. (Illustrate by pushing the long pendulum too rapidly and the short pendulum too slowly.)

Lesson Notes

1. Be sure your students adjust both pendulums to the same length. This ensures a complete energy transfer from one pendulum to the other in step 3. If they have different lengths (and therefore, frequencies), this energy transfer will be incomplete, similar in form to step 5.

2. Looping each pendulum around the rubber band prevents it from sliding down the thread. This may not be necessary if the pendulums are spaced at least 1 paper length apart. Friction between the rubber band and thread usually (though not always) is sufficient to hold the rubber band in place.

3. This transfer of energy from one pendulum to the other will astonish and delight your students. Describing its motion provides good exercise in making accurate written observations.

5. The illustration suggests that one pendulum be made about twice as long as the other. If length differences are exaggerated to a ratio of 1 to 4 , the short pendulum will swing exactly twice as fast as the long one. This results in almost no energy being transferred at all! Every time the longer pendulum swings in phase one way, it swings exactly out of phase going back again.

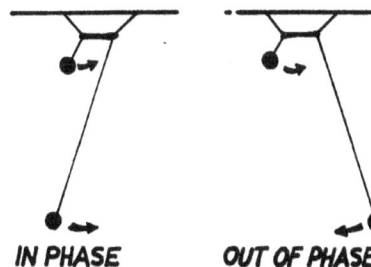

IN PHASE   **OUT OF PHASE**

The net result is that only the pendulum originally placed in motion swings at all. The other remains nearly motionless.

Check Point

3. The pendulums shift their energy of motion from one to the other. As one slows to a stop, the other reaches maximum energy. Then the process reverses.

4. A high rubber band transfers the energy between pendulums rather slowly. As you move it down, these energy transfers become more rapid.

5. No. When you set one pendulum in motion, it will not be able to transfer much energy to the other. Because it has a different frequency (different length), it swings out of step with the other.

6. The pendulum you swing first transfers only part of its energy before gaining it back again. It slows down and speeds up while the other, originally at rest, starts and stops.

NAME: CLASS:

COIN HYPNOSIS

1 Hang 2 coin pendulums from your table so they just touch. Neither should be able to touch a table leg.

COINS JUST TOUCH.

CAN'T TOUCH TABLE LEGS.

2 Color the tape on *one* coin with your pencil or pen. Leave the other coin alone.

3 While 1 pendulum hangs still, swing the other around it.

Give a HARD, WIDE swing!

4 Keep your *eye* on the *same* coin. Describe the energy transfer you see.

5 Lengthen 1 pendulum so it almost touches the floor. Leave the other half as long.

Almost to the floor.

a. *Predict* if either pendulum can transfer *all* its energy to the other. Explain.

b. Swing the *short* pendulum around the *long* one. Explain if your prediction was correct.

...AROUND and AROUND, and...

c. Swing the *long* pendulum around the *short* one. Explain if your prediction was correct.

Objective

To observe energy transfers between pendulums that wind around each other. To predict how this motion changes as you make one pendulum longer than the other.

Introduction

None required.

Lesson Notes

1. Make both pendulums as long as possible, but not so long that they touch a table leg, or other obstruction, when swinging one around the other.

Step 5 requires that one of these pendulums be extended. Leaving excess thread on one of them now means you won't need to bother making a longer one later.

3. The harder you swing the coin, the more fascinating its motion and dramatic the result.

5. The previous activity provides the basis for predicting energy transfers here. When pendulums coupled by the rubber band had different lengths, one kept swinging while the other stopped and started. That is, the energy transfer was incomplete.

After you lengthen one pendulum, be sure to moderate your swing when you start the longer one revolving around the shorter one. Otherwise, you will likely hit the coin against a table leg or other obstruction.

Extension

Lengthen one pendulum so it almost touches the floor. Shorten the other to one-fourth this length.

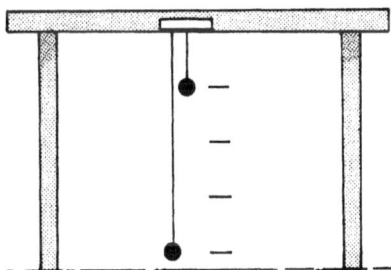

a. Experiment to see how these two pendulums transfer energy.

(There is almost no energy transfer! Either pendulum can swing around its stationary neighbor, moving it hardly at all. Because the short pendulum swings exactly twice as fast as the longer one. They move in, then out, of phase with every revolution.)

b. Swing one pendulum perpendicular to the other so they cross paths, back and forth, without touching or winding around each other.

(Simply release both pendulums together. Because the short one swings twice as fast as the long one, they'll not collide in the middle for a long while.)

Check Point

4. One coin makes smaller and smaller circles until it finally stops; the other coin moves into wider and wider circles until it reaches a maximum. Then the pendulums transfer energy back again. This back and forth energy transfer repeats faster and faster as the two pendulums continue to wind together.

5a. Energy transfers will be incomplete. Because pendulums of unequal length have unequal frequencies, neither pendulum can deliver energy to the other that is totally in phase.

b. As predicted, the short pendulum transferred only part of its energy to the long one. It continually swings around the longer, losing some energy then gaining it back, while the longer pendulum starts to swing, then stops again.

c. Again, as predicted, the long pendulum transferred only part of its energy to the short one. It continually swings around the shorter, losing some energy then gaining it back, while the shorter pendulum starts and stops.

B. PENDULUM MATHEMATICS

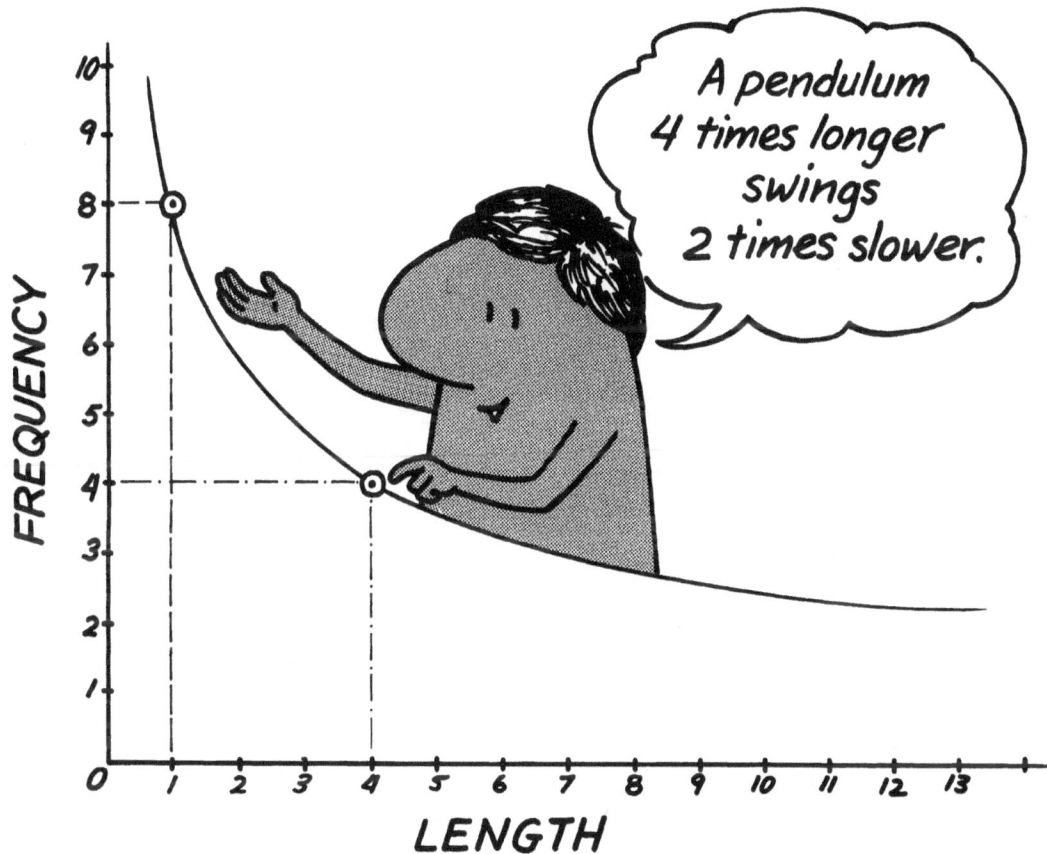

Your students have already determined pendulum frequency by counting cycles over one-minute intervals. They know qualitatively that short pendulums swing at higher frequencies than long pendulums. But they have not yet expressed this relationship in quantitative terms. Until now.

These experiments help students describe pendulum motion mathematically, to apply quantitative relationships in predictive ways. Knowing that pendulums swing to the tune of inverse squares, your students can predict new frequencies and lengths from experimental values already known. They'll graph the inverse square relationship between length and frequency into graceful curves. They'll even use pendulum frequencies to find an experimental value for the $\sqrt{2}$.

All this predictive number power does not come without effort. Your students will find this series of activities more difficult than last. They'll have to work harder and you'll need to explain things in greater detail. Nevertheless, with effort they will succeed.

Mathematics is the language of science. As your students experiment and reason in this new language, they'll gradually become fluent, and grow in scientific literacy.

EVALUATION

Each question evaluates a single activity from PENDULUM MATHEMATICS as numbered. Use any combination to frame a formal exam or an informal review: Copy these questions on your blackboard, construct your own ditto master, or photocopy the questions while masking out the rest of the page. Evaluate in ways that suit your own teaching style, enabling your students to learn and enjoy science.

Questions

B-1 A pendulum makes 20 swings in 15 seconds.

    a. How many cycles does it make in 15 seconds?
    b. What is the frequency of the pendulum in cycles per minute?

B-2

A   B   C   D   E

1 cm
3 cm
5 cm
9 cm
12 cm

Fill in each blank with the correct letter.

a. Pendulum _____ swings 3 times faster than pendulum _____ .
b. Pendulum _____ swings 2 times faster than pendulum _____ .

B-3
Draw a smooth graph line through these data points.

FREQUENCY (1 sq = 25 c/m)
LENGTH (1 sq. = 10 cm)

a. According to your graph, what is the frequency of a pendulum that is 35 cm long?

b. According to your graph, how long is a pendulum with a frequency of 37.5 c/m?

B-4
A chain with 2 paper clips has a frequency of 147 c/m. Find the frequency of a chain with 18 paper clips. Show your work.

B-5

FREQUENCY
LENGTH

This graph illustrates how the frequency of a paper clip pendulum changes with length. Why is the curve steep at the top and shallow at the bottom?

B-6
A pendulum 100 cm long swings 30.4 c/m. Another pendulum 200 cm long swings 21.5 c/m. Use this information to find the square root of 2. Show your math.

B-7
Pretend you are rich enough to build a pendulum in any manner you wish. Design one that comes close to perpetual motion — once you start it, it never stops.

Answers

B-1
a. 10 cycles in 15 seconds
b. 40 cycles/minute
B-2
a. Pendulum A swings 3 times faster than pendulum D.
b. Pendulum B swings 2 times faster than pendulum E.
B-3
(Your students should draw a smooth curve that connects *most* of the points. The graph line should not cut through any of the circles.)
a. 50 c/m
b. 64 cm
B-4
An 18-link chain is 9 times longer, so it swings 3 times slower: 147 c/m ÷ 3 = 49 c/m

B-5
Short pendulums are more sensitive to changes in length than long pendulums. The shorter they get, the faster the frequency rises (steep curve). By contrast, the longer pendulums get, the slower the frequency decreases (shallow curve).

B-6
$$\sqrt{2} = \frac{30.4 \text{ c/m}}{21.5 \text{ c/m}} = 1.414\ldots$$

B-7
(Accept any reasonable answer. Here is one example.) Build a long, heavy pendulum with a knife-edge pivot. Enclose the whole system in a large bell jar and pump out all the air.

SEQUENCING

PENDULUM MATHEMATICS logically follows from A. Because of its quantitative nature, it is more challenging. If your class is already familiar with pendulum variables and can accurately measure frequency in cycles per minute, skip A and begin here.

Related Activities: A---**B**

MATERIALS

Here is everything your students will use for the next 7 activities on PENDULUM MATHEMATICS. Materials printed in normal type are part of the core 15-things-in-a-box inventory that support all 100 activities. Materials printed in *italics* are additional local materials that you provide or ask your students to bring from home. Pencil and paper are already assumed and therefore unlisted. Each item is numbered with the activity where it is first used.

(B-1) Thread.
(B-1) Scissors.
(B-1) Paper clips of uniform size and weight.
(B-1) Masking tape.
(B-1) A *wall clock* with second hand sweep. Your students can also use wristwatches, if they have them.
(B-6) *Small coins* of equal size.
(B-6) *Hand calculators* (optional).

FURTHER STUDY

Use problems like these plus "extension" ideas in PENDULUM MATHEMATICS to lead your students beyond worksheet activity into original research and investigation. Each discovery leads to more questions, deeper questions, better questions than these. Answering them is what good science is all about.

We measured pendulums in cycles per minute. How else do scientists measure pendulums? Find out about "Hertz" units. How is a pendulum's "period" different from frequency?

How fast must something vibrate before we can hear the sound it makes? Can things vibrate too fast to hear? Write a report about sound.

Name other things that move in cycles. The ocean tides, for example, move in much slower cycles, while a vibrating guitar string moves much faster. Write down as many processes that you can think of. Place all your events on a time line in the correct order.

COUNTING CYCLES

1 Tie thread to 3 paper clips. . .

...then loosely fold masking tape over each thread so it slides back and forth.

2 Put each paper clip in the middle of its pattern. Slide each tape to the exact distance shown:

◄— *PIVOT TO CENTER* —►

7cm

6cm

5cm

Write the correct length on each tape.

3 Count cycles for each pendulum during 15 seconds, 30 seconds and 60 seconds. Fill in the table.

ONE CYCLE = TWO SWINGS

length	15 sec.	30 sec.	60 sec.
7 cm			
6 cm			
5 cm			

4 To find frequency in cycles/minute (c/m) do you have to count for a whole minute? Describe a shortcut method.

5 If you counted the cycles of a 5 cm pendulum without stopping, how high would you count after one week?
Show your math.

...mumble, mumble...

Objective

To find the frequency of a 5 cm, 6 cm and 7 cm pendulum. To discover a short-cut method for determining frequency.

Introduction

Make a pendulum. Use it to teach your students how to *accurately* count cycles. Be sure to cover these points: (1) Avoid the practice of counting "one cycle" after only the first swing. Demonstrate how you must wait one *full* cycle before counting "one." If you release the pendulum's paper clip near your body, for example, you must not count "one" until the paper clip swings back toward you a second time. (2) If time runs out between cycles, report frequency values to the nearest 1/2 cycle. Don't round off.

Lesson Notes

2. Frequency in short pendulums is extremely sensitive to changes in length, much more so than in long pendulums. Because these 3 pendulums are relatively short, their frequencies cannot be accurately determined unless they are constructed to the exact lengths specified; hence, the actual-size patterns. Longer pendulums in later activities, because they are less sensitive, can be measured with a simple ruler.

To use these patterns, first slide the masking tape on each pendulum near its paper clip bob. Then match the center of this paper clip to the center of the pattern. (It doesn't matter if your particular paper clips are larger or smaller than the pattern, as long as the centers match.) Finally pull the thread through the tape until the pivot positions match.

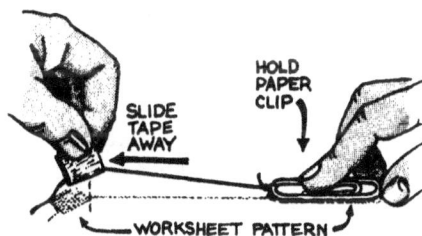

If you can't depend on your students to check their own accuracy, require them to show their pendulums to you before proceeding to step 3. Any pendulum that is off by more than a millimeter should be readjusted.

3. Some students will likely notice that each column is a simple multiple of the next. To finish the table they need only multiply. Others may complete the entire table by experiment. Either method is fine. Step 4 prompts all students to recognize the multiple relationship.

4. When you multiply a measurement, you also multiply its uncertainty. A frequency that is off by one cycle over fifteen seconds, for example, will be off by four cycles when multiplied to a whole minute. It's quicker and easier to count cycles over a short time interval, but it's also less accurate.

On the other hand, it's easy to become distracted and lose track of long counts that last a whole minute. Counting short, high-frequency pendulums over full-minute intervals is especially difficult. In general, we find that counting cycles for 30 seconds, then multiplying by 2, yields the most consistent and accurate frequency results. Encourage your students to do the same.

Check Point

3.

length	15 sec.	30 sec.	60 sec.
7 cm	28	56½	113
6 cm	30½	61	122
5 cm	33½	67	134

4. No. You can count cycles over shorter time intervals and then multiply. A 15 second count multiplied by 4, or a 30 second count multiplied by 2 is also equivalent to a full minute count.

5. A 5 cm pendulum swings at 134 cycles per minute:

134 c/m x 60 m/hr x 24 hr/day x 7 days/wk = 1,350,720 c/wk

PENDULUM SQUARES

1 Pendulums swing to a beautiful number pattern.
Find this pattern to complete the table.

LENGTH	**1**	**4** TIMES LONGER	**9** TIMES LONGER	**16** TIMES LONGER	TIMES LONGER	TIMES LONGER	TIMES LONGER	TIMES LONGER	TIMES LONGER	TIMES LONGER
FREQUENCY	**1**	**2** TIMES SLOWER	**3** TIMES SLOWER	**4** TIMES SLOWER	TIMES SLOWER	TIMES SLOWER	TIMES SLOWER	TIMES SLOWER	TIMES SLOWER	TIMES SLOWER

2 **a.** You know the frequency of a 7 cm pendulum. Use this table to *predict* the frequency of a pendulum that is 4 time longer (28 cm).

— 28 cm —

b. Make a pendulum and test your prediction:

Measure with this ruler.

3 **a.** You know the frequency of a 6 cm pendulum. Use the table to *predict* the frequency of a 54 cm pendulum.

b. Make a pendulum and test your prediction:

4 **a.** You know the frequency of a 5 cm pendulum. Use the table to *predict* the frequency of a 80 cm pendulum.

b. Make a pendulum and test your prediction:

5 A 1 cm pendulum swings at 300 c/m. Knowing this, use the table above to calculate other frequencies:

LENGTH	**1** cm	**4** cm	**9** cm	**16** cm
FREQUENCY	**300** c/m	c/m	c/m	c/m

Given... 1 cm = 300 c/m

For your math:

cm: 20 15 10 5 0

Objective

To discover a squared relationship between pendulum frequency and length. To apply this relationship in a predictive way.

Introduction

Copy this table on your blackboard. Draw a short pendulum next to it. Because it is only 2 cm long, it manages to swing at a very rapid 212 cycles each minute.

length	1	4 TIMES LONGER	9 TIMES LONGER	16 TIMES LONGER
freq.	1	2 TIMES SLOWER	3 TIMES SLOWER	4 TIMES SLOWER

freq. = 212 c/m

Use this information to calculate with your class the frequency of . . .
... an 8 cm pendulum. (106 c/m)
... an 18 cm pendulum. (71 c/m)
... a 32 cm pendulum. (53 c/m)

Confirm that your are, after all, predicting accurate frequency values. Build a 32 cm paper clip pendulum to see if it does indeed swing 53 cycles back and forth in 1 minute. (It does.)

Can your class now appreciate the power of pendulum mathematics and apply it in a predictive way? If so, they are ready to continue.

Lesson Notes

1. Each term in the frequency sequence, 1, 2, 3, etc, is squared to produce the length sequence:

$$1^2, 2^2, 3^2, 4^2, 5^2, 6^2, 7^2, 8^2, 9^2, 10^2, \ldots$$

Others may build this length sequence by adding successive differences:

1, 4, 9, 16, 25, 36, 49, 64, 81, 100, . . .
3 5 7 9 11 13 15 17 19

2-4. Students must *first* predict what the pendulum frequencies should be, based on the table. Then they should experiment to see if they are correct.

Make these pendulums in the usual manner: tie a paper clip to some thread; fold masking tape over the thread; pull this thread through the masking tape to make a pendulum of any desired length. If excess thread is used, about 85 cm, a single pendulum system will adjust to all three lengths.

Measure each pendulum by repeating ruler lengths at the side of the page. Because these pendulums are relatively long, measuring accuracy is not as critical. Lengths that vary up to a centimeter cause only minor variations in frequency.

5. You can improvise a 1 cm pendulum by attaching a tiny ball of clay or other small weight to a short piece of thread. To *roughly* estimate its frequency, count cycles for *one* second and multiply by 60. (A 1 cm pendulum swings about 5 cycles a second, or 300 cycles/minute.)

Check Point

1.

LENGTH	1	4 TIMES LONGER	9 TIMES LONGER	16 TIMES LONGER	25 TIMES LONGER	36 TIMES LONGER	49 TIMES LONGER	64 TIMES LONGER	81 TIMES LONGER	100 TIMES LONGER
FREQUENCY	1	2 TIMES SLOWER	3 TIMES SLOWER	4 TIMES SLOWER	5 TIMES SLOWER	6 TIMES SLOWER	7 TIMES SLOWER	8 TIMES SLOWER	9 TIMES SLOWER	10 TIMES SLOWER

2a. A 7 cm pendulum swings 113 c/m. A 28 cm pendulums is 4 times longer so it swings 2 times slower.
(113/2 = 56.5 c/m)

b. (Students should compare their experimental value with this calculated value.)

3a. A 6 cm pendulum swings 122 c/m. A 54 cm pendulums is 9 times longer so it swings 3 times slower.
(122/3 = 40.7 c/m)

b. (Students should compare their experimental value with this calculated value.)

4a. A 5 cm pendulum swings 134 c/m. An 80 cm pendulum is 16 times longer so it swings 4 times slower.
(134/4 = 33.5 c/m)

b. (Students should compare their experimental value with this calculated value.)

5. A 1 cm pendulum swings 300 c/m. Thus a . . .
... 4 cm pendulum swings twice as slow.
(300/2 = 150 c/m)

... 9 cm pendulum swings 3 times as slow.
(300/3 = 100 c/m)

... 16 cm pendulum swings 4 times as slow.
(300/4 = 75 c/m)

GRAPH YOUR PENDULUMS

1 Fill in this table using frequency values from the last 2 activities.

2 Plot and circle each point on the graph.

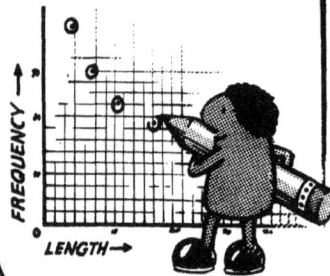

3 Draw a smooth graph line along the points.

RIGHT: smooth curve　　　WRONG: wavy curves

THE LINE MIGHT NOT TOUCH ALL THE CIRCLES!

DON'T DRAW INSIDE CIRCLES

FREQUENCY →
LENGTH →

B-2
B-1

CUTOUTS
B-3

LENGTH (cm)	1	4	5	6	7	9	16	28	54	80
FREQUENCY (c/m)										

FREQUENCY (c/m)

300

200

100
80
60
40
20

0　5　10　15　20　25　30　35　40　45　50　55　60　65　70　75　80　85

LENGTH (cm)

4 Read your graph to fill in each value.

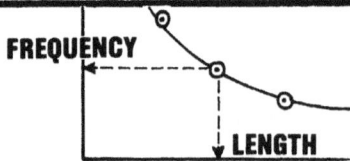

FREQUENCY

LENGTH

LENGTH (cm)	6		18		35		70
FREQUENCY (c/m)		90		60		40	

5 What does your graph tell you about pendulum frequency and length?

Objective

To graph pendulum frequency as a function of length. To read and interpret the resulting graph.

Introduction

If your students don't understand how to graph, begin with a class discussion. Copy these coordinates plus the data table on your blackboard. Demonstrate how to plot the first data point in the table (10,95). Locate 10 cm on the horizontal axis, and 95 c/m on the vertical axis. Then plot the point where perpendiculars from these two locations intersect. Ask volunteers to plot other coordinates in a similar manner.

Length (cm)	10	15	22	28
Freq. (c/m)	95	78	61	57

Next, show how to connect these points with a smooth line. To keep the graph line from obscuring hard data points, draw a circle around each one. The graph line stops at the perimeter of each circle, then continues on the other side.

Lesson Notes

1. Your students have already found frequencies for all pendulum lengths listed in the data table. Nothing more is required here than to transfer this data from previous worksheets.

3. Pendulum frequencies change according to a well-defined trend. If the data points don't all seem to follow this trend, it's not the fault of the pendulums. (They always swing just like they are supposed to!) Rather, experimental error has scattered the points to some extent. To compensate, draw the best possible *smooth* line through *most* of the points. Attempting to connect points that are scattered by error will result in a graph that looks more like an economic indicator than a pendulum curve.

4. Graphs are great predicting tools. Each point on the graph line pairs a specific length to a specific frequency. Given one variable, a set of perpendicular lines always leads to the second. Where you land is what you predict!

Check Point

1-3.

LENGTH (cm)	1	4	5	6	7	9	16	28	54	80
FREQUENCY (c/m)	300	150	134	122	113	100	75	57	41	34

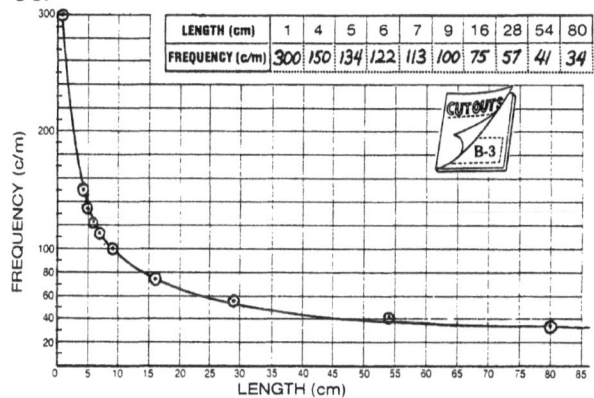

4. Here is one result. Student answers may vary by several points, depending on how they place the graph line.

LENGTH (cm)	6	12	18	23	35	47	70
FREQUENCY (c/m)	122	90	67	60	48	40	35

5. (It is easy to see that the graph line goes down. The implications are obvious. But what about the curve in this graph line? What does that mean? Challenge your more capable students to look beyond the direction of the line to notice its shape as well.)

The frequency of a pendulum decreases with length. The curve in the graph line further indicates that the *rate* of frequency change also decreases with length. This means that the frequency of short pendulums is much more sensitive to length changes than the frequency of long pendulums.

PAPER CLIP MATHEMATICS

1 Link 8 paper clips to form a chain. Hang it from a ninth clip that has an arm bent out.

Use the ninth clip as a handle.

Find the frequency of this 8-clip chain:

2 Keep adding to the same chain to make it 12 clips long.

Don't count the clip you hold.

Then make a 16-clip chain.

a. Find the frequency of this 12-clip chain:

b. Find the frequency of this 16-clip chain:

3 Paper clip chains swing to the same rule of squares as thread pendulums. Complete this number pattern as before.

LENGTH	1	4 TIMES SHORTER	9 TIMES SHORTER	16 TIMES SHORTER	TIMES SHORTER	TIMES SHORTER	TIMES SHORTER	TIMES SHORTER	TIMES SHORTER	TIMES SHORTER
FREQUENCY	1	2 TIMES FASTER	3 TIMES FASTER	4 TIMES FASTER	TIMES FASTER	TIMES FASTER	TIMES FASTER	TIMES FASTER	TIMES FASTER	TIMES FASTER

4 Use the information above to complete this table.

You know these values...

... calculate the rest based on these.

LENGTH	8 clips	2 clips	12 clips	3 clips	16 clips	1 clips	4 clips	9 clips
FREQUENCY	c/m	c/m	c/m	c/m	c/m	c/m	c/m	c/m

For your math:

Objective

To mathematically predict the frequency of paper-clip chains. To confirm these predictions by experiment.

Introduction

This activity is logically equivalent to B-2. Students who understood the squared relationship between frequency and length for thread pendulums can apply the same logic to these paper clip chains. This does not imply that everyone will have an easy time completing this worksheet. Problems that require quantitative reasoning are always extra challenging. Be ready, therefore to provide extra help with step 4.

Lesson Notes

Ever since there were paper clips, there have been students who love linking them together into long chains. Now they can do so without getting into trouble! Paper clip chains provide a new context in which to review, once again, the mathematical relationship between pendulum length and frequency.

1-2. The clip you hold functions only as a support. As such, it must not be counted as a link in the swinging chain. Bending out the arm of the support clip helps remind students that it serves this unique function.

3. Notice that this table describes *shorter* pendulums swinging *faster*. The table in B-2 describes longer pendulums swinging slower. Either way, the numbers are the same.

4. The quantitative logic required to fill in the first square of the first table goes like this: By experiment, a chain with 8 swinging clips makes about 73 cycles in one minute. A two-clip chain is 4 times shorter than 8 clips. According to the pendulum squares relationship, it therefore swings 2 times faster. Multiplying 73 c/m by 2 yields 146 c/m. You can check the accuracy of this prediction by experiment. Hook 2 clips together, swing them from a third, and count the swings for ten seconds. The result, multiplied by 6, should be close to 146 c/m, the calculated value.

The last entry in this table is tricky. (That's why it's last.) The frequency for a 9 clip chain is based on 1 clip, not 16 clips. Being 9 times *longer*, it must swing 3 times *slower*. (In this one particular case, the "longer/slower" wording in table B-2 is more appropriate.) Students who find all this too confusing might simply link 9 paper clips together and find the experimental value.

Extension

You know the frequency of a 1-clip chain. What is the frequency of a chain with 100 clips?

The key to solving this problem mathematically is to find a standard of comparison. Here you should compare 100 clips with 1 clip. The chain is 100 times longer, so it should swing 10 times slower. Dividing the frequency for 1 clip by 10 yields a predicted frequency of 20.8 cycles.

Care to test this answer? It makes an entertaining class demonstration. Have five groups of students each make a chain 20 clips long, then string them all together. Remember to add on one extra clip, number 101, to use as a support. Using standard clips, the chain will extend over 3 meters (about 10 feet). You'll need to hang the chain over a balcony, down a stairwell, or out a second-story window, wherever you can clear the ground.

Do your really want to stretch your students' imaginations? Ask them how many clips it would take to build a giant pendulum that requires a full minute to complete just one cycle.

(Here is the answer: We know that 1 paper clip swings 208 cycles per minute. A giant pendulum 208 times slower swings the required 1 cycle per minute. Applying the pendulum square rule, it should be 208^2 times longer than 1 clip, that is, 43,264 clips long! At 100 clips per box it would require about 433 boxes to build this pendulum. Knowing that 1 box of 100 clips stretches a little over 3 meters, this pendulum would reach about 1300 meters — well over a kilometer.)

Imagine you are riding on the end of this chain. What a ride!

Check Point

Because paper clip brands vary in size, answers that your students get may differ somewhat from these. Values should be internally consistent, however. If your paper clips are smaller than ours, expect frequencies that are always higher than these figures (or lower for larger clips). Encourage your students to develop a number sense: Does the answer look right? Is it what one would estimate?

1. An 8 clip chain swings at 73 c/m.

2a. A 12 clip chain swings at 60 c/m.

   b. A 16 clip chain swings at 52 c/m.

3.

LENGTH	1	4 TIMES SHORTER	9 TIMES SHORTER	16 TIMES SHORTER	25 TIMES SHORTER	36 TIMES SHORTER	49 TIMES SHORTER	64 TIMES SHORTER	81 TIMES SHORTER	100 TIMES SHORTER
FREQUENCY	1	2 TIMES FASTER	3 TIMES FASTER	4 TIMES FASTER	5 TIMES FASTER	6 TIMES FASTER	7 TIMES FASTER	8 TIMES FASTER	9 TIMES FASTER	10 TIMES FASTER

4.

LENGTH	8 CHDS	2 CHDS	12 CHDS	3 CHDS	16 CHDS	1 CHDS	4 CHDS	9 CHDS
FREQUENCY	73 c/m	146 c/m	60 c/m	120 c/m	52 c/m	208 c/m	104 c/m	69 c/m

NAME: CLASS:

Pendulum Mathematics **B-5**

CHAIN GRAPH

1 Fill in the table with paper clip frequencies you already know. Experiment to find the rest.

REMEMBER— DON'T COUNT THE CLIP YOU HOLD!

This is a **5-CLIP CHAIN**

2 Plot and circle each point.

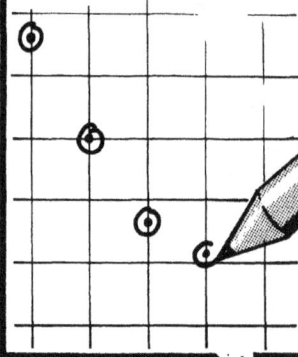

3 Draw a smooth graph line along the points.

DON'T DRAW INSIDE THE CIRCLES

CUTOUTS
B-5

LENGTH (clips)	FREQ. (c/m)
1	
2	
3	
4	
5	
6	
7	
8	
9	
12	
16	

FREQUENCY (c/m): 220 200 180 160 140 120 100 80 60 40 20 0

LENGTH (paper clips): 1 2 3 4 5 6 7 8 9 10 11 12 13 14 15 16 17

4 Your graph line curves down steeply, then gets more and more shallow. Explain what this means.

Objective

To graph how the frequency of a paper-clip chain varies with length.

Introduction

This activity is similar to B-3, the previous graphing activity. Remind students to use the same graphing procedures as before: draw the best possible *smooth* line through the data points; enclose each point in a circle, keeping it clear of the graph line.

Lesson Notes

1. Your students have already determined the frequencies of all but three paper clip chains: 5 clips, 6 clips, and 7 clips.

2-3. Paper-clip chains swing by the same laws of motion as thread-and-paper-clip pendulums do. It is not surprising, therefore, that the two graphs look similar. There are, nevertheless, important differences to notice. Observe that the horizontal scale is measured in paper-clip lengths rather than more conventional centimeters. Notice as well that the chain graph is displaced higher up the frequency scale. The distance from the pivot to the center of mass in a chain is always less than in a thread-and-paper-clip pendulum of equal length because the chain has a higher center of mass.

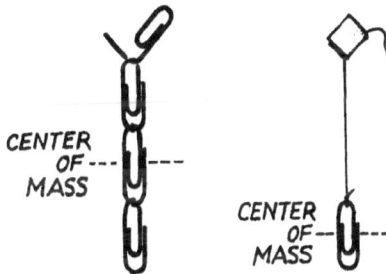

CENTER OF MASS

CENTER OF MASS

Extension

Do your students enjoy graphing? Consider plotting paper clips chains and bunches on the *same* graph.

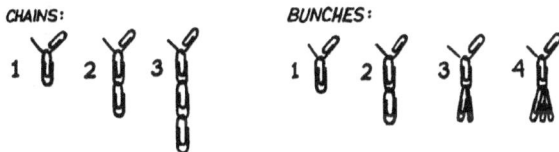

CHAINS: 1 2 3

BUNCHES: 1 2 3 4

As usual, plot number of clips on the horizontal axis against frequency on the vertical axis. The resulting graph dramatically illustrates a principle first discovered in activity A-3: Increasing the mass of a pendulum's bob does not change frequency; length is the only variable that matters.

Check Point

1-3.

LENGTH (clips)	FREQ. (c/m)
1	208
2	146
3	120
4	104
5	92
6	84
7	78
8	73
9	69
12	60
16	52

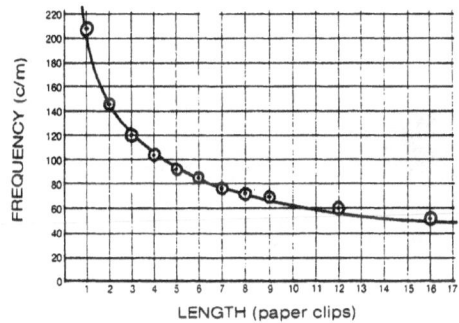

4. As you add more and more clips to the chain (and thus move down the horizontal axis), the rate of change in the frequency decreases. The addition of just another paper clip to so many already on the chain causes only a small drop in frequency. This makes the curve flatten out.

NAME: CLASS:

SQUARE ROOT OF TWO

1 What number multiplied by itself makes 2?

ANSWER:

$$\sqrt{2} = 1.41421\,3$$

THIS IS A NON-REPEATING DECIMAL!

5329...

You can find this number using pendulums!

2 Tape a coin to some thread. Tie a knot at the other end.

Hold it by the KNOT.

Find its frequency.

3 Now make your pendulum exactly half as long.

AT HALF-LENGTH, THE KNOT SHOULD HANG TO THE MIDDLE OF THE COIN.

a. Find its frequency.

4 Here's another way to find the square root of 2. Make a large square with equal sides, then draw its diagonal.

EQUAL SIDES

DIAGONAL

a. Find the length. . .

. . .of the diagonal.

. . .of the side.

b. Divide the frequencies to find the square root of 2.

$$\sqrt{2} = \frac{\text{FREQUENCY of HALF PENDULUM}}{\text{FREQUENCY of LONG PENDULUM}}$$

b. Now divide the diagonal by the side.

$$\sqrt{2} = \frac{\text{DIAGONAL}}{\text{SIDE}}$$

c. How close did you come to 1.414?

c. How close did you come to 1.414?

METRIC RULER   0 cm 1 2 3 4 5 6 7 8 9 10 11 12 13 14 15 16 17 18 19 20

Objective

To calculate the square root of two, first by using pendulum ratios, then by using simple geometry.

Introduction

Your students should be familiar with the concept of taking square roots. Here are some problems to try as a class exercise. Begin with perfect squares:

$$\sqrt{4} = 2, \quad \sqrt{25} = 5, \quad \sqrt{144} = 12, \text{ etc.}$$

Once your students catch on, ask them the really tough question: What is $\sqrt{2}$? Using a calculator, you can show by trial and error that it is a number very close to the value given in step 1.

> 2 x 2 = 4, so it's less than 2;
> 1 x 1 = 1, so it's more than 1;
> 1.5 x 1.5 = 2.25, so it's less than 1.5;
> 1.4 x 1.4 = 1.96, so it's greater than 1.4;
> 1.42 x 1.42 = 2.0164, so it's less than 1.42;
> 1.414213 x 1.414213 = 1.9999984:
> Close, but not exact.

You'll never find an *exact* decimal equivalent to this irrational number. The ancient Greeks tried and tried, without success. Messy, non-repeating decimals did not sit well with their philosophy of a universe that ran on simple ratios and whole numbers.

Lesson Notes

2-3. This knot is important. It defines the pivot position on both pendulums. When swinging the full pendulum, students actually hold this knot between their fingernails — no higher, no lower. Then they use it again to locate the pivot of the half-size pendulum.

3-4. Converting fractions to decimals may confuse some students. They easily forget which part of the fraction (numerator or denominator?) divides into the other. Don't tell them the answer. Rather, suggest this useful problem solving strategy:

To figure out a complex problem, try substituting a similar problem with simple numbers. Understand how numbers in this easy problem are related, then apply the pattern you discover to the more complex problem:

You can, of course, omit all arithmetic by providing calculators. But if your students need long division practice, by all means, let them practice!

4. There are many ways to construct a large square. You can fold paper into a square, draw a square on graph paper, or use a triangle and ruler. Let your students figure out their own methods.

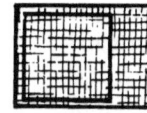

FOLDED-PAPER SQUARE   GRAPH PAPER SQUARE

Extension

Use paper clips to calculate the square root of 3. Multiply your root to find out how close you came.

You can find $\sqrt{3}$ or any other square root by using pendulum ratios, just as before:

$$\sqrt{3} = \frac{\text{freq. of a 3-clip chain}}{\text{freq. of a 9-clip chain}}$$

$$\sqrt{5} = \frac{\text{freq. of a 5-clip chain}}{\text{freq. of a 25-clip chain}}, \text{ etc.}$$

Be sure to hold your chain by 1 extra clip.

Check Point

2. Answers will vary.

3. Fractions will vary depending on the particular length of each pendulum chosen. All answers, of course, should approximate $\sqrt{2}$ =1.414.

4. Again, fractions will vary depending on the particular size of each square chosen. But the correct answer must again approximate 1.414.

NAME: CLASS:

HALF LIFE

1 Suspend a straightened paper clip just off the floor. Hook it through an end loop that you tied around your pencil.

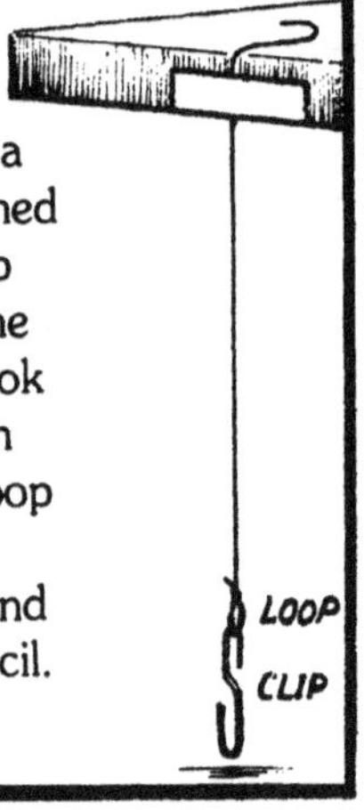

LOOP

CLIP

2 Cut out the amplitude chart. Place it on the floor so the paper clip *almost* touches the center X.

3 Release the clip at the full amplitude line. Count how many cycles it makes before it decays to the half amplitude line.

3, 4, 5 ...

FULL AMPLITUDE

HALF AMPLITUDE

Write your results in the table next to 1.

4 Continue adding clips to the loop and counting cycles. Graph your results.

BOB MASS (clips)	DECAY TIME (cycles)
0	
1	
2	
3	
4	
6	
8	
10	

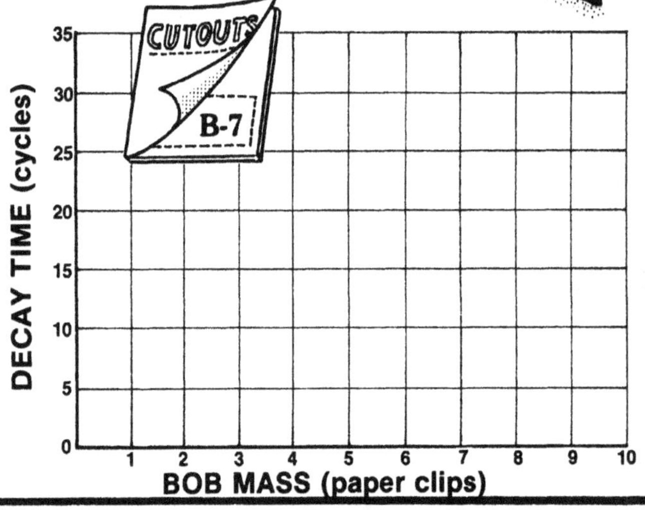

CUTOUTS B-7

DECAY TIME (cycles) — 0, 5, 10, 15, 20, 25, 30, 35

BOB MASS (paper clips) — 1, 2, 3, 4, 5, 6, 7, 8, 9, 10

5 **a.** Why does the amplitude decay?

b. Interpret the shape of your graph line.

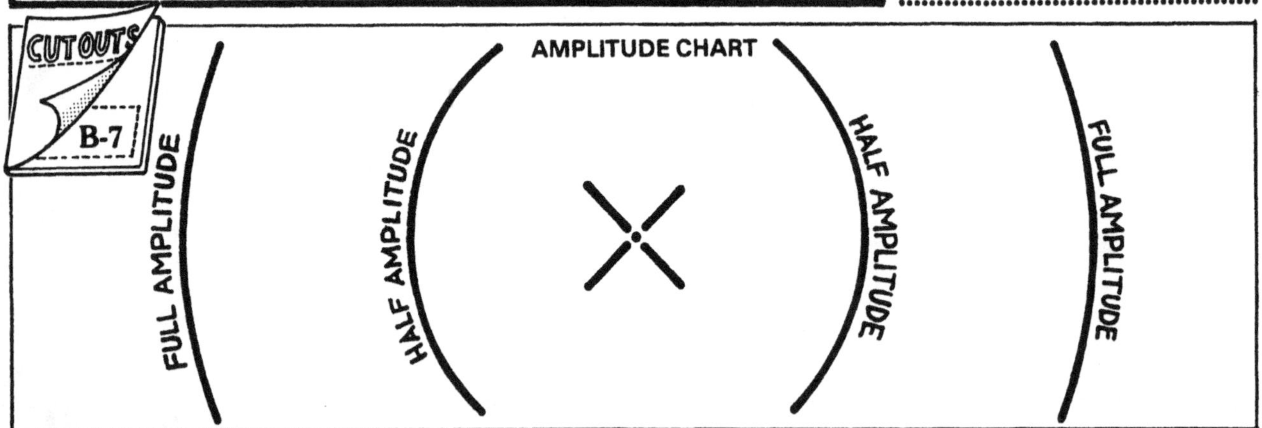

CUTOUTS B-7

AMPLITUDE CHART

FULL AMPLITUDE   HALF AMPLITUDE   X   HALF AMPLITUDE   FULL AMPLITUDE

Objective

To graph how the weight of the bob affects amplitude decay in a pendulum.

Introduction

Write the word "decay" on your blackboard and discuss its meaning. Within the context of this activity it means to gradually decline or slow down.

This is the first activity in the entire program that actually directs students to cut something out (step 2). Impress upon students (as dramatically as you can) that scissors must *never* be applied to the pages of any *Student Reference Book*, now or at any time in the future. These books are for *reference only*. Use instead the amplitude charts that are printed in the consumable *Student Cutouts Booklets* under reference number B-7.

Lesson Notes

1. After tying the loop of thread, slide it back off your pencil. Cut off any excess that hangs past the knot.

2-3. Placing the straightened paper clip close to the chart reduces parallax. You can observe with greater accuracy the pendulum bob's actual position relative to lines on the chart.

3-4. Notice that the data table begins with zero clips, not one. After filling in the value for 1 paper clip (from step 3) students must go back and decide what they should enter next to 0 clips. The question is this: how many times will thread swing back to the half-amplitude line with no clips attached? The answer is 0 cycles. Try it and see. With no weight attached, the thread drifts slowly down to the center X and stays there. Filling in the rest of the table, from 2 paper clips up to 10 is simply a matter of adding them one at a time to the loop in the thread.

Notice how the straightened clip hangs down further than the ones you add. It functions as a marker, making it easier to define exactly *where* the cluster should be released over the full amplitude line, and *when* it passes over the half amplitude line for the last time. As you add clips, forming a larger and larger cluster, be sure to add them first to one side of this marker clip, then to the other. This keeps the marker on center where it should be. Even if the cluster rotates, it won't displace the marker clip from its center position.

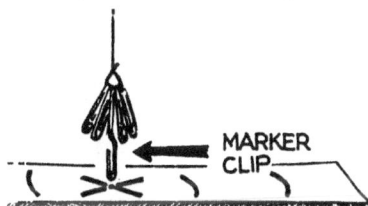

MARKER CLIP

As you add paper clips to the bob, the amplitude decays more slowly, and thus approaches the half-amplitude line more gradually. This makes it increasingly difficult to tell exactly when (at what cycle) the bob crosses the half-amplitude line for the last time. You can expect, therefore, wider margins of error (an increased scattering of points) with increased bob weight. Nevertheless, the overall trend in the graph line is still clear. It climbs steeply from zero, then gradually flattens.

It is interesting to speculate how this experiment might change if it were carried out on the airless moon. Because there is almost no atmosphere to slow it down, the pendulum would swing much longer before decaying to the half amplitude line. You would thus need to count huge numbers of cycles and find the graph line rising almost vertically off the page (unless you changed the scale). You would also need to count more slowly. Pendulums swing at lower frequencies on the moon because of its reduced gravitational pull. Due to internal resistance in the thread to continuous bending (near the pivot), each pendulum would eventually decay past half amplitude, and the experiment would *eventually* grind to an end. Nothing lasts forever, not even on the moon.

Check Point

4. Results depend on the size (mass) of the paper clips you use.

BOB MASS (clips)	DECAY TIME (cycles)
0	0
1	7
2	11
3	14
4	17
6	22
8	27
10	31

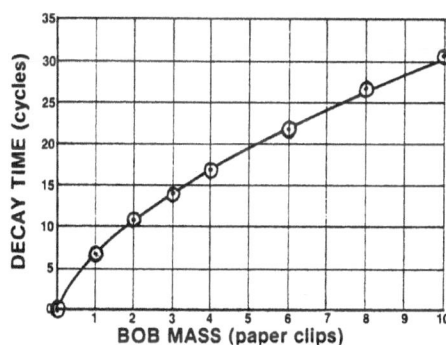

5. The pendulum continually pushes air out of its way as it swings. This air resistance takes energy away from the pendulum, causing its amplitude to decay.

6. The graph line curves up into a gentle arc that gradually flattens. Adding more paper clips to the bob enables the pendulum to store increasing amounts of energy and thus swing through more cycles before decaying to the half amplitude line. As the mass of the bob increases, it becomes increasingly insensitive to the addition of one more clip.

C. BALANCE BEAMS

Hanging paper clips from numbered positions on a beam and making them balance is great fun, a wonderful way to learn math: your students will add and subtract whole numbers, write equations, manipulate equalities and inequalities and learn about symmetry.

Until now math balances have usually been hand-crafted from wood, and therefore very expensive. Not any more. TOPS math balances are so cheap that all students in your class can build their own. Fold them from just a single sheet of paper. Cut and tape as directed. Use paper clips for weights. Then step aside and let the learning begin.

This series of activities requires little or no initial teacher input. As students progress from worksheet to worksheet, engrossed in the trial and error of creative play, they will learn all they need to know about the physics of balance beams and appreciate their mathematical beauty.

─── EVALUATION ───

Each question evaluates a single activity from BALANCE BEAMS as numbered. Use any combination to frame a formal exam or an informal review: Copy these questions on your blackboard, construct your own ditto master, or photocopy the questions while masking out the rest of the page. Evaluate in ways that suit your own teaching style, enabling your students to learn and enjoy science.

Questions

C-3
Use your math balance (if necessary) to decide if this beam balances.

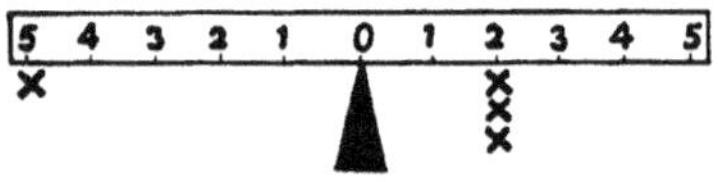

C-6
Draw 4 different ways to make this beam balance by adding 3 paper clips to the right arm.

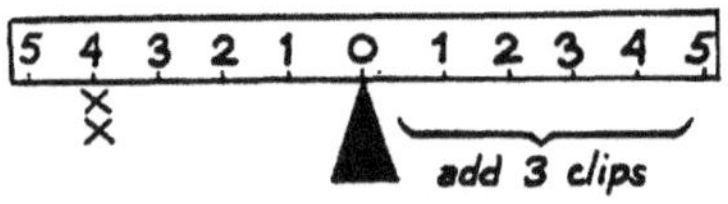

add 3 clips

C-4
Draw 2 more X's to make this beam balance. Choose your positions carefully so you can write a different equation in each box.

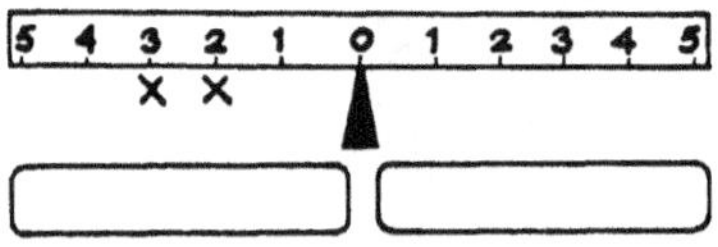

C-5
Draw X's under positions 3 and 4 to make this beam balance. Write a multiplication equation in each box.

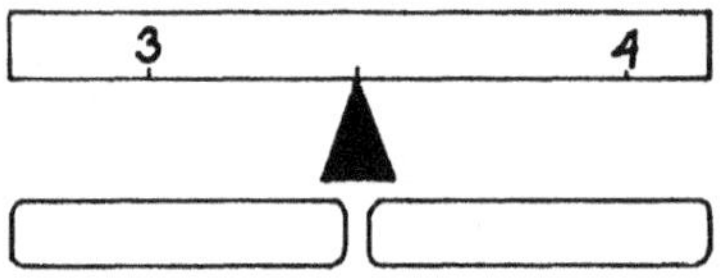

C-7
Write an equation in each box. Does this beam balance?

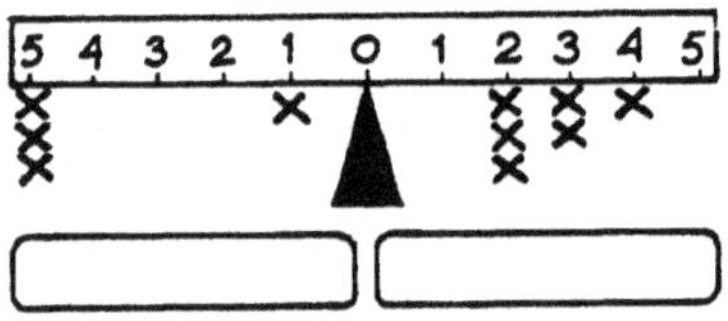

C-8
Find the beam that balances. Show your math.

a.

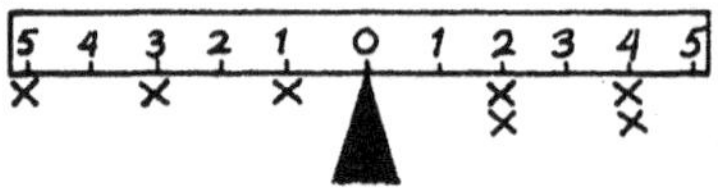

b.

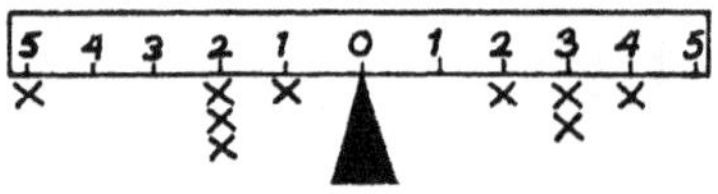

c.

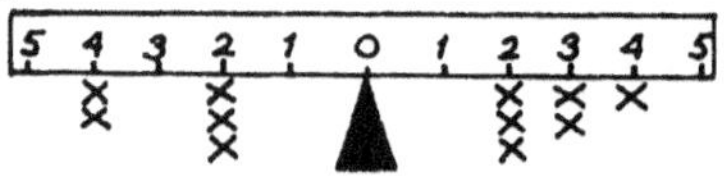

C-9
Find the beam that tilts left . Show your math.

a.

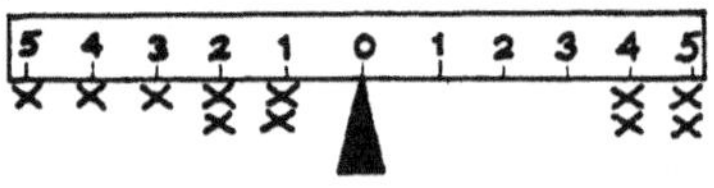

b.

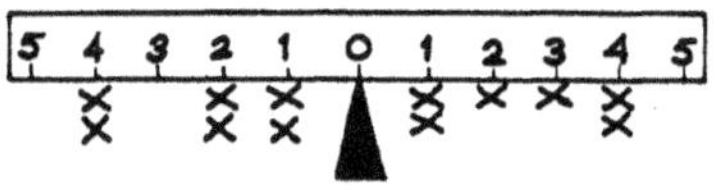

c.

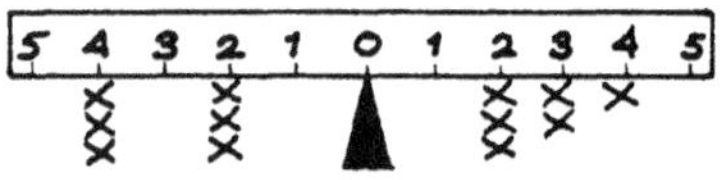

Answers

C-3
No. The beam tilts down to the right.

C-4
3 + 2 = 5   1 + 4 = 5

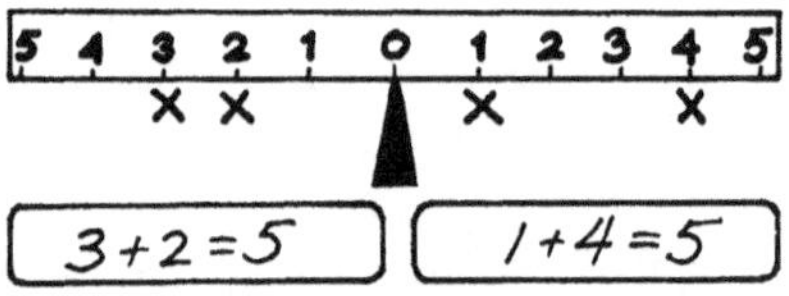

C-5
3 x 4 = 12   4 x 3 = 12

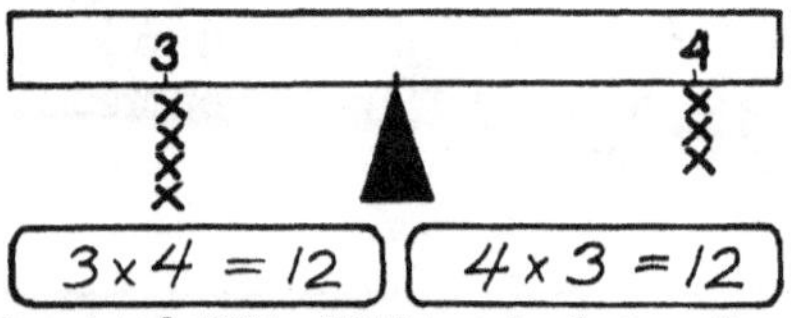

C-6

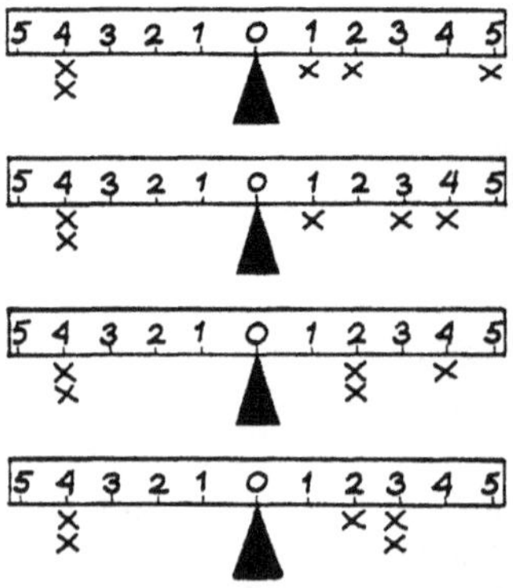

C-7
Yes. 5×3 + 1 = 16   2×3 + 3×2 + 4 = 16

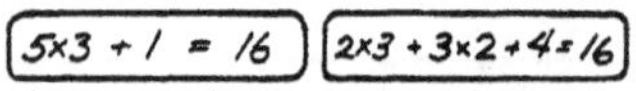

C-8
a. L: 5 + 3 + 1 = 9
   R: 2 x 2 + 4 x 2 = 12

b. L: 5 + 2 x 3 + 1 = 12
   R: 2 + 3 x 2 + 4 = 12

c. L: 4 x 2 + 2 x 3 = 14
   R: 2 x 3 + 3 x 2 + 4 = 16

Beam "b" balances.

C-9
a. L: 5 + 4 + 3 + 2 x 2 + 2 = 18
   R: 4 x 2 + 5 x 2 = 18

b. L: 4 x 2 + 2 x 2 + 2 = 14
   R: 2 + 2 + 3 + 4 x 2 = 15

c. L: 4 x 3 + 2 x 3 = 18
   R: 2 x 3 + 3 x 2 + 4 = 16

Beam "c" tilts left.

SEQUENCING

BALANCE BEAMS may be studied at any time. Because the series is quantitative, yet relatively easy, schedule it sooner rather than later. The balance base developed in **C** will also be used in D. It is convenient, therefore, to schedule **C** first to avoid detouring back to the assembly directions in C-2.

Related Activities: **C---D**

MATERIALS

Here is everything your students will use for the next 9 activities on BALANCE BEAMS. Materials printed in normal type are part of the core 15-things-in-a-box inventory that support all 100 activities. Materials printed in *italics* are additional local materials that you provide or ask your students to bring from home. Pencil and paper are already assumed and therefore unlisted. Each item is numbered with the activity where it is first used.

(C-1) Scissors.
(C-1) Clear tape.
(C-2) Masking tape.
(C-2) Spring action clothespins.
(C-2) *Medium-sized cans.*
(C-2) Straight pins.
(C-2) Paper clips of uniform size and weight.

FURTHER STUDY

Use problems like these plus "extension" ideas in BALANCE BEAMS to lead your students beyond worksheet activity into original research and investigation. Each discovery leads to more questions, deeper questions, better questions than these. Answering them is what good science is all about.

Read about "machines" and "work". Explain how your balance beam operates like a lever.

Construct mobiles from string, washers, straws and tape. Examine the mathematics of why these mobiles balance. Hold an art contest. Be creative.

Consider non-mechanical concepts of balancing. How do you balance your checkbook? A mathematical equation? What do we mean by an ecological balance? How does the inner ear help you maintain a physiological sense of balance? Are you eating a balanced diet?

BUILD A MATH BALANCE (1)

1 START HERE:
Carefully trim this paper along the outer dotted line.

2 Fold it in half exactly along the center line.

3 Cut *between* the black tabs to remove all 9 grey areas.

DON'T cut through the black tabs.

4 Fold up *both* bottom edges to just touch the base of each black tab. Don't cover the tabs.

FOLD BOTH EDGES UP TOGETHER

5 Fold up the bottom edge a third time. Remember to just touch the base of each black tab above. Don't cover the tabs.

This is your *third* fold....

6 Tape shut between the tabs with clear tape.

7 Open each tab to form a loop.

A PENCIL HELPS.

8 Write your name here.

FOLD

CENTER

5 4 3 2 1 0 1 2 3 4 5

Copyright © 1988 by TOPS Learning Systems. Reproduction limited to personal classroom use only.

Objective

To fold a paper beam to use in a math balance.

Introduction

This worksheet itself actually folds into a balance beam as students complete instructions 1-8 written upon it. Those who take the time to read directions and understand illustrations *before* folding and cutting should experience little difficulty.

Unfortunately, some students will not take this necessary time. In step 3, for example, they can cut off black tabs quicker than you can say "follow directions". Worksheets can, of course, always be repaired with clear tape and students can always start over. But it is easier to head off problems before they occur. Prevention is easier than cure.

The prevention you need to apply in this activity is to build the balance yourself, before your students try. This will familiarize you with the directions, and provide a model for your students to follow. Then, just before students begin work on their own, review steps 1 through 4 together. Take a student's worksheet (from the cutout booklet, of course), then actually cut and fold while students watch. You don't need to remove all the grey areas in step 3. Cutting out just one is sufficient to illustrate where to make cuts in this critical step. Then demonstrate the correct fold in step 4, and hand the worksheet back to the student you borrowed it from. This will give everyone a clear idea about how to proceed.

Lesson Notes

2. When folded together along the guide line, the top and bottom edges of the paper don't quite meet. These edges will eventually "creep" together as the beam is folded over in step 4, and again in step 5.

Some students may fold the paper inward and thereby cover the directions. The illustration directs students to fold their worksheets outward.

3. Students are directed here to cut out the grey areas *between* black tabs. Watch out for those that cut *through* them instead. These black tabs will be cut more uniform and even if students start at the bottom of each tab then cut upward into the apex of each grey area.

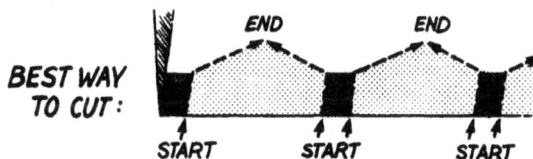

BEST WAY TO CUT:

END   END

START   START   START

This is easier than cutting around the grey areas in one continuous motion.

MORE DIFFICULT WAY TO CUT:

START ..... END   START ...... END

4. *Both* bottom edges fold up as illustrated, not just one.

8. A completed balance beam should look something like this.

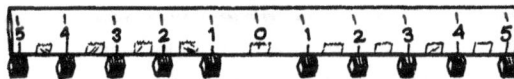

The beams your students make may not appear so well put together. But don't despair. Despite bad cuts and crooked folds, most beams still work reasonably well.

Check Point

Is the paper well folded so the edges match more or less evenly? Are the tabs properly cut and pushed to open loops? Tabs that are sliced off or cut too narrow should be patched back together with clear tape and then recut. Tabs that are cut too wide (with grey showing along the cut edges) should be trimmed.

NAME: CLASS:

BUILD A MATH BALANCE (2)

1 Fold masking tape over the ends of a clothespin. Pinch the ends flat.

AS WIDE AS A PAPER CLIP.

2 Cut out a narrow strip from the center of the tape.

LOOKS LIKE EARS!

CUT TO THE WOOD

3 Clamp the clothespin to the side of a tin can like this.

4 Lay your beam across a clothespin. Then push a straight pin through the *exact* center of the crossmarks.

TAPE helps protect your finger!

CENTER

5 Rest your beam on top of the clothespin between the ears. (It doesn't need to balance level.)

EARS

5 4 3 2 1   1 2 3 4 5

6 Bend out 2 paper clips just a little. . .

. . .then hang them from the outside loops like this.

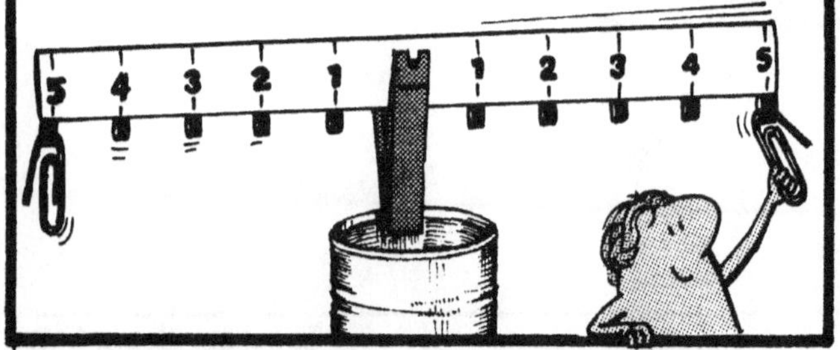

5 4 3 2 1   1 2 3 4 5

7 Make a "rider". Roll up some tape so just one end is sticky.

LEAVE A STICKY TAB

FOLD A "HANDLE"

8 Make the beam balance level by adding this rider to the lighter (higher) side.

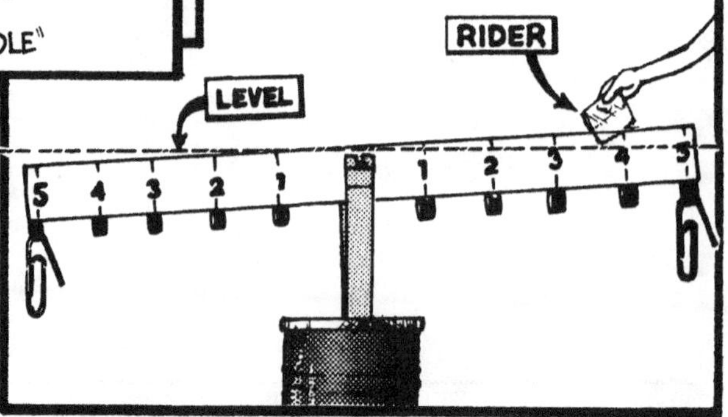

RIDER

LEVEL

5 4 3 2 1   1 2 3 4 5

Objective

To complete construction of the math balance.

Introduction

None required.

Lesson Notes

1. This tape must not extend too far beyond the ends of the clothespin. Otherwise it will curl and interfere with the free motion of the beam in between. If tape does extend beyond the width of a paper clip, simply cut off the excess.

2. If your scissors are blunt, you may need to cut down each side of the slot, then scrape away the narrow piece of tape still attached at the bottom. Remove this piece completely, so the pivot pin can rest directly on solid wood in step 5.

3. Soda bottles also serve as balance bases. Some bottles may have openings with just the right diameter, enabling a clothespin to fit snugly inside. If the mouth of the bottle is too large, you can make it narrower by sticking tape around the inside of the rim.

TAPE

— MOUTH TOO LARGE —

GOOD FIT

Soda cans also work nicely, if they are equipped with pull tabs still connected to the can. Simply push the tab into a vertical position, then attach the clothespin.

If you elect to use tin cans or soda cans as the balance base, you can dramatically increase their overall stability by filling each one-third full with gravel, sand, or even dirt.

4. Notice that this clothespin is not the same as the one used in step 3. This one doesn't form part of the balance. It functions instead as a pin cushion, providing an open space underneath so the pin can completely pierce the paper. Students who attempt this step while holding the beam in their fingers risk getting poked with the pin.

It is important to push the pin through the exact center of the cross mark. Younger students may require special help to accurately position the pin.

5. Students should not expect their beams to rest level. Not yet. In the process of making them, they have likely used extra tape or cut away more paper on one side of their beams than on the other. Steps 8 will surely set things right.

6. A paper clip place on each end of the beam (they must weigh the same!), lowers its overall center of gravity. This makes the beam less prone to drift unpredictably from side to side. Even though the beam still doesn't rest level, it should now return quickly to the same equilibrium position whenever you push it off balance.

8. Move the rider right or left until the beam balances level. If the rider is too light to overcome any tilt, make a heavier one.

This process of leveling the beam by shifting the rider is called "centering". Normally a balance should center empty, reaching a level equilibrium with no paper clips at all. Adding a paper clip to each side of the beam makes this leveling process easier to accomplish, but also reduces the overall sensitivity of the beam.

Check Point

Is the pin pushed through the beam exactly on center? Does the *empty* beam swing freely, returning each time to a level position?

If the beam drifts slowly from side to side, instead of centering into a level position, the pivot is too low relative to the beam's overall center of gravity. In the process of cutting and folding this particular beam, an unusual amount of weight remains on the high side. To fix this, simply reposition the pin a little higher up the vertical mark.

In general, the beam becomes more stable as you place the pin higher up the vertical marks and more unstable (but more sensitive) as you place the pin lower. The intersection of these two marks is a compromise between stability and sensitivity.

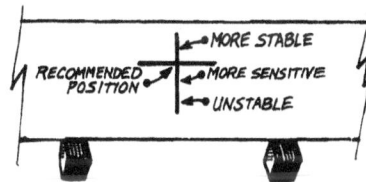

MORE STABLE
MORE SENSITIVE
RECOMMENDED POSITION
UNSTABLE

PAPER CLIP BALANCING

1 Pull out the arms on 9 paper clips just a little.

PULL IT OUT THIS FAR ONLY.

2 Start with a *level* beam. Then add paper clips and make it balance level again.

5 4 3

3 Draw 8 *different* ways to make your beam balance.

Here's one way…

5 4 3 2 1 1 2 3 4 5

a. 5 4 3 2 1 0 1 2 3 4 5

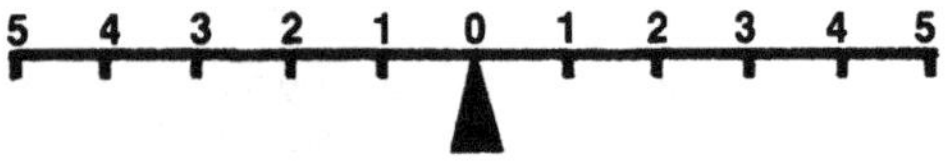

b. 5 4 3 2 1 0 1 2 3 4 5

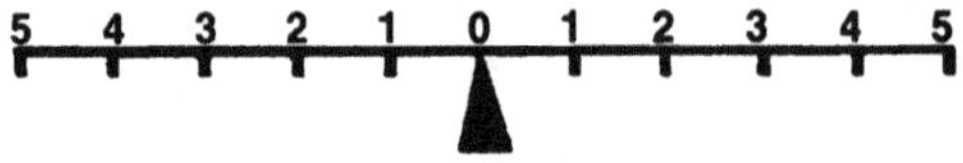

c. 5 4 3 2 1 0 1 2 3 4 5

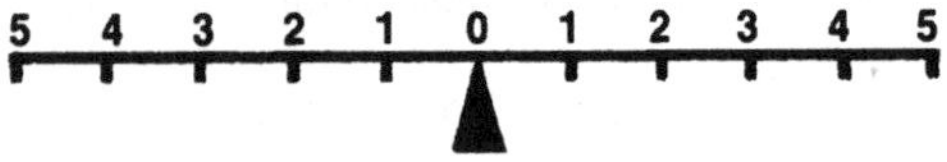

d. 5 4 3 2 1 0 1 2 3 4 5

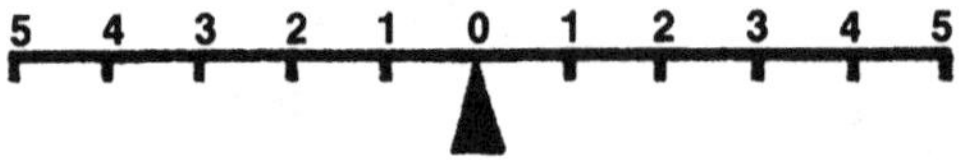

e. 5 4 3 2 1 0 1 2 3 4 5

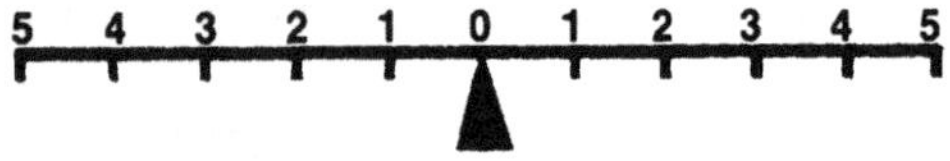

f. 5 4 3 2 1 0 1 2 3 4 5

g. 5 4 3 2 1 0 1 2 3 4 5

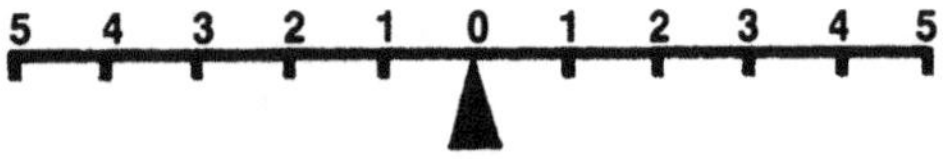

h. 5 4 3 2 1 0 1 2 3 4 5

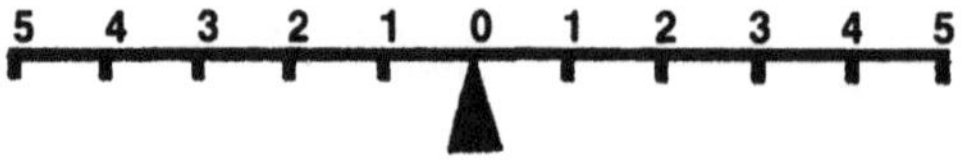

Objective

To get acquainted with a math beam. To diagram various ways that paper clips can balance on the beam.

Introduction

Start with an *empty* off-center balance. (Move the rider if you have to.) Then demonstrate how to re-center the beam by shifting the rider right or left until it balances level again. Emphasize how important this procedure is. In the routine course of handling, balance beams may sometimes begin to tilt off center. The starting point for any balancing activity, therefore, must be a level beam. Good results depend on it.

Now ask a class volunteer to add just 3 paper clips to the beam, so that 2 on one side balance just 1 on the other side. Once accomplished, challenge your whole class to find other wonderful ways to balance paper clips.

Lesson Notes

1. All paper clips used for balancing must have uniform size and weight. Otherwise the balance will not function in a simple mathematical way. Isolate odd brands of paper clips and banish them from your classroom.

Tell students to pull out the "arms" of the paper clips just a little. Students tend to exaggerate this operation, bending them so much that they can't be used for other purposes.

It is important to limit students to 9 clips per balance in this worksheet. They can use additional clips in later balancing activities, but for now 9 is an ideal number. Students have enough clips to easily create many different balance variations, but not enough to overload the balance. When the beam gets too heavy with paper clips, it loses sensitivity. It is difficult to determine if the paper clips exactly balance or just almost balance.

2-3. Only one paper clip needs to be looped through any black tab. Additional clips can be added in chains or clusters.

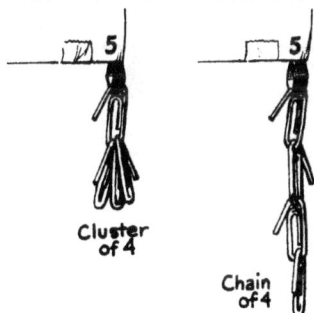

Cluster of 4

Chain of 4

3. Students typically begin this exercise by balancing equal numbers of clips equal distances from the pivot.

This is correct, of course, but somewhat boring, if they repeat this pattern throughout. The introduction encourages students to balance *unequal* paper clip distributions as well.

Variations on the "X" symbol used to represent paper clips are fine. Artists in your class may prefer to draw real paper clips. Do not, however, allow students to substitute actual numbers. If three paper clips belong under a specific number, students should write "XXX" but not "3."

Your more advanced students may discover, with great triumph, that their math balances can add and multiply! Others will require more time to make this discovery. Please be patient. Don't rush in and try to explain everything. *All* students can experience the joy of discovery learning and gain self confidence in their abilities to learn *if* you provide the necessary space and time.

These math balances will last longer if students are responsible to maintain their own. To this end, assign a specific place in your room where students can store their assembled balances and paper clips.

Check Point

(There are many possible paper clip configurations. Answers that appear suspect should be reconfirmed on a balanced beam.)

NAME: _____   CLASS: _____

BALANCE ADDITION

1 Be sure your beam balances level.

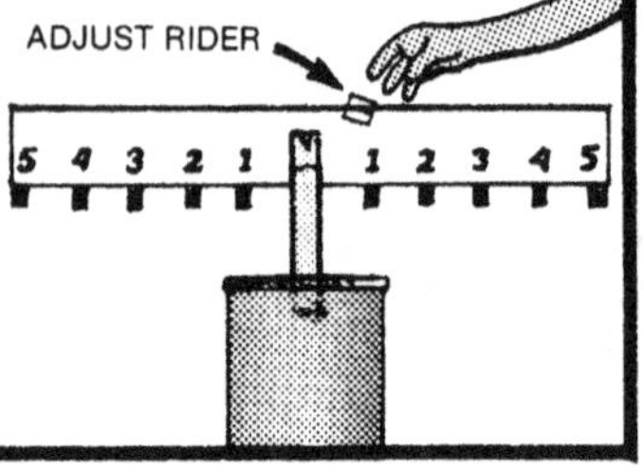

ADJUST RIDER

5 4 3 2 1   1 2 3 4 5

2 Now draw all the *different* ways you can make 2 paper clips on the left balance just 1 on the right.

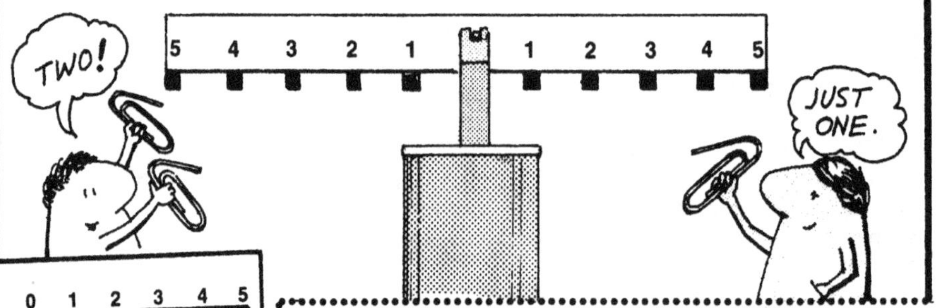

TWO!

5 4 3 2 1   1 2 3 4 5

JUST ONE.

a. EXAMPLE

5 4 3 2 1 0 1 2 3 4 5

X
X   X

Write an equation under each balance.

$1 + 1 = 2$

b.

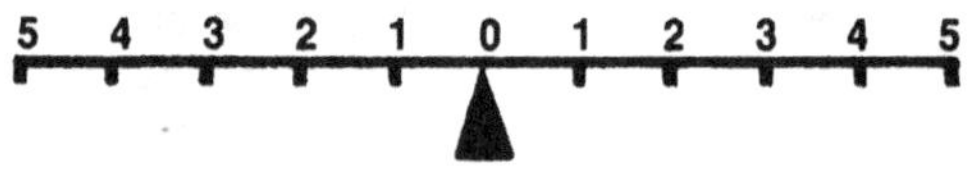

5 4 3 2 1 0 1 2 3 4 5

c.

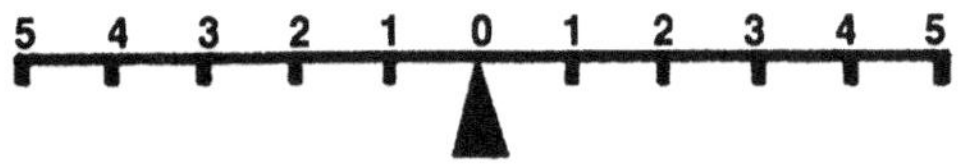

5 4 3 2 1 0 1 2 3 4 5

d.

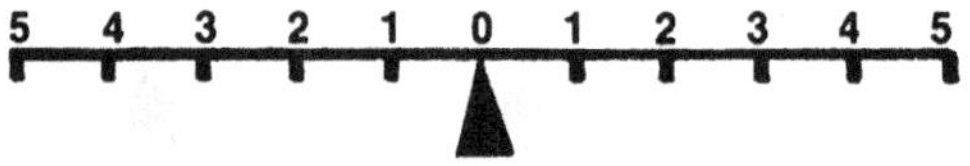

5 4 3 2 1 0 1 2 3 4 5

e.

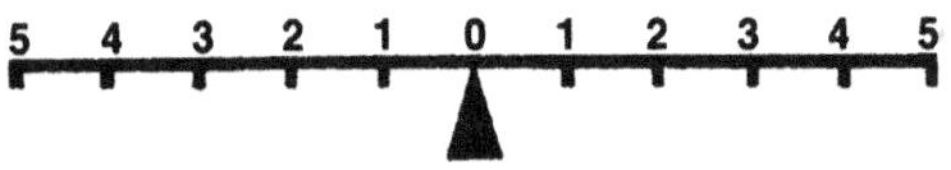

5 4 3 2 1 0 1 2 3 4 5

f.

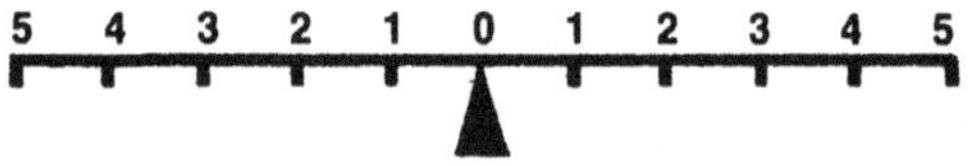

5 4 3 2 1 0 1 2 3 4 5

Objective

To understand that paper clips add up to equal sums on each arm of a balanced beam.

Introduction

This activity, plus the 5 that follow all present balancing puzzles for your students to think about and solve. While all TOPS activities support individualized activity, this sequence is especially well-suited to independent study. If you have always wondered about self-paced study, but have never given it a try, here is a low-risk opportunity for you to see how it works.

Tell your class to complete the rest of these balancing activities on their own. They must remember three things: (1) Make sure the balances are centered *before* beginning each new activity. (2) Read the directions and study the examples *before* asking the teacher for help. (3) Bring each completed activity to the teacher for a check-point approval *before* starting the next.

Lesson Notes

There are 6 unique ways to make 1 paper clip balance 2.

Check Point

Accept these answers in any order. Watch out for duplicate paper clip configurations.

2b. 2c. 2d. 2e. 2f.

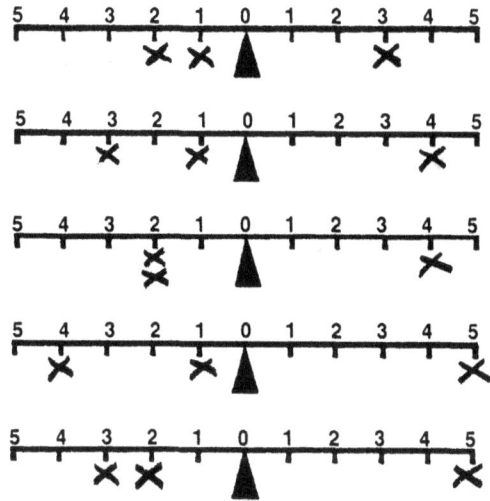

NAME: ______________________ CLASS: ______________

BALANCE MULTIPLICATION

1 Start with an empty balance beam.

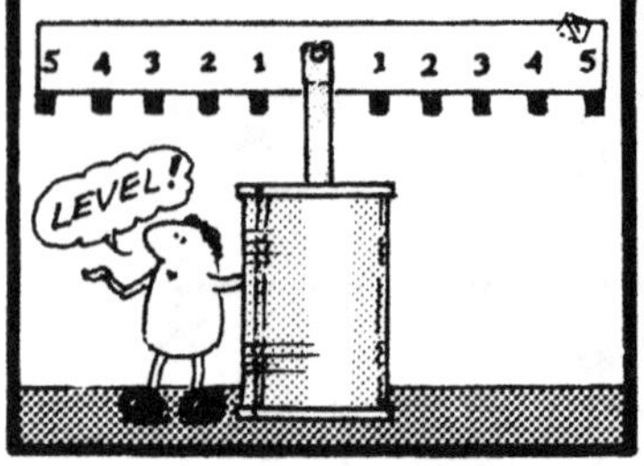

5 4 3 2 1 1 2 3 4 5
LEVEL!

2 Make your beam balance by adding clips to only *one* tab on each side. The tabs *must* be different.

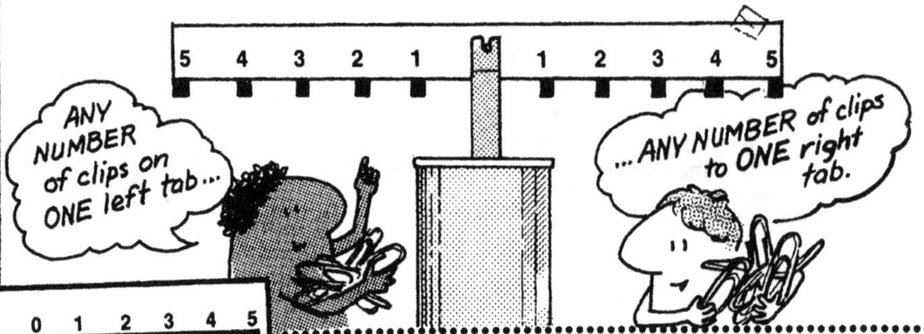

5 4 3 2 1 1 2 3 4 5
ANY NUMBER of clips on ONE left tab...
...ANY NUMBER of clips to ONE right tab.

a.

EXAMPLE
5 4 3 2 1 0 1 2 3 4 5
2×3=6
3×2=6
Write an equation under each side of the balance.

b.

5 4 3 2 1 0 1 2 3 4 5

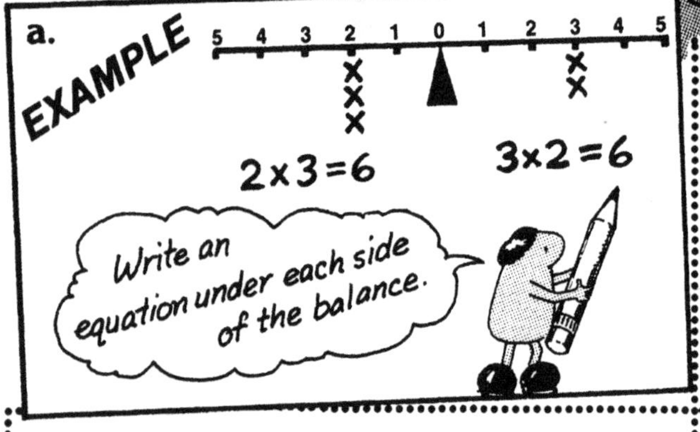

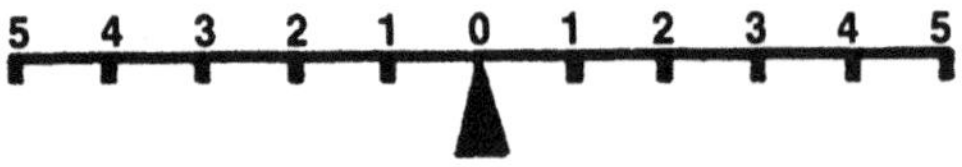

c.

5 4 3 2 1 0 1 2 3 4 5

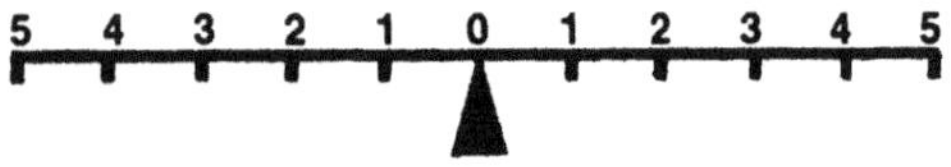

d.

5 4 3 2 1 0 1 2 3 4 5

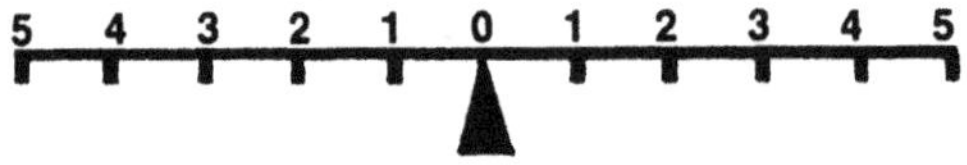

e.

5 4 3 2 1 0 1 2 3 4 5

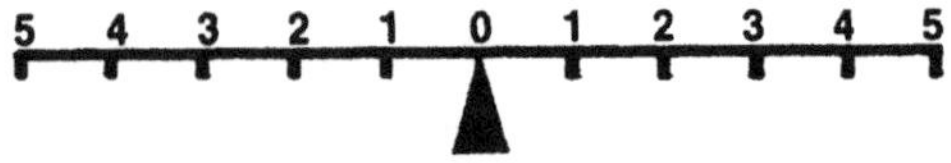

f.

5 4 3 2 1 0 1 2 3 4 5

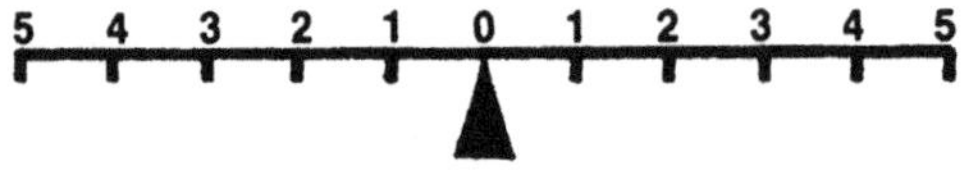

Objective

To understand that paper clips multiply to equal products on each arm of a balanced beam.

Introduction

None required.

Lesson Notes

Choose any tab on one side of the beam. Select a different tab on the other side. If you add paper clips *one at a time* to these tabs, always to the side that is higher, the beam *must*, eventually, reach a state of balance. Consider, for example, what happens if you always add clips, one at a time, to the higher of positions 5 or 3:

START:

BALANCE:

Your students, of course, won't be as methodical. Adding paper clips in bunches, it is quite possible they may overlook the first multiple (in this case 5x3 = 3x5) and report a higher multiple instead (5x6 = 3x10).

Check Point

Accept answers like these in any order.
2a. 2b. 2c. 2d. 2e. 2f.

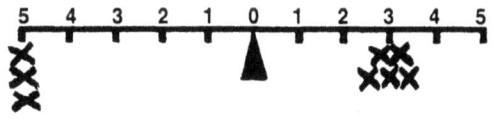

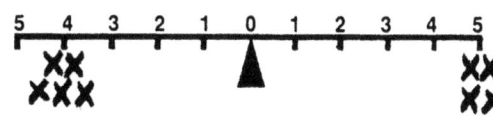

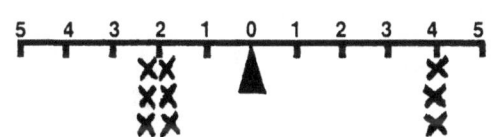

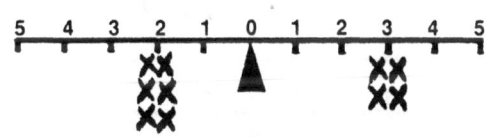

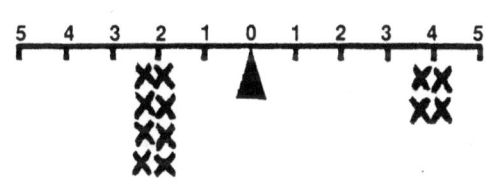

NAME: CLASS:

BALANCE PUZZLES

1 Show 4 *different* ways to balance this beam by adding only 3 paper clips to the right side.

2 CLIPS ALWAYS HERE

Always put 3 CLIPS on this side.

Show that each side always equals 10.

a. EXAMPLE

$5 \times 2 = 10$ $3 \times 2 + 4 = 10$

b.

$5 \times 2 = 10$

c.

$5 \times 2 = 10$

d.

$5 \times 2 = 10$

2 Repeat with 4 clips. Use only the first 4 tabs on the right side.

Always put 4 CLIPS on this side.

NO CLIPS HERE

a.

$5 \times 2 = 10$

b.

$5 \times 2 = 10$

c.

$5 \times 2 = 10$

d.

$5 \times 2 = 10$

Objective

To understand the mathematics of balancing. To gain further experience with balance beams.

Introduction

None required.

Lesson Notes

As students understand the mathematics of the balance beam, they tend to shift from physical activity (actually placing paper clips on the beam) to mental activity (writing number combinations that add up to the desired result). This is ideal. As a result of concrete manipulations, these students have advanced to a higher level of mental abstraction.

1. You can help those who experience difficulty in finding all four balance combinations by introducing the concept of symmetry. Notice how each new solution is derived from the one above by symmetry moves. Once the beam is placed in a state of balance, then clips that are moved in equal but opposite directions will maintain that balance.

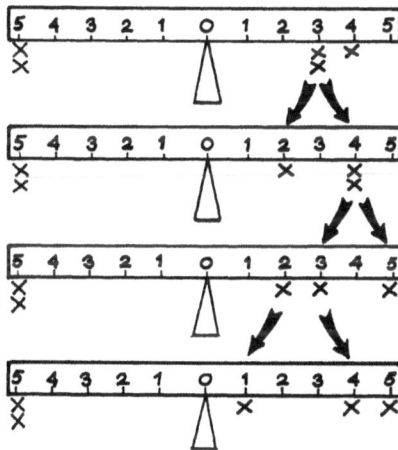

2. Notice that solutions to the last group of problems are restricted to the first 4 beam positions on the right side. Position number 5 is off limits.

Check Point

Accept these answers in any order.

1b. 1c. 1d.

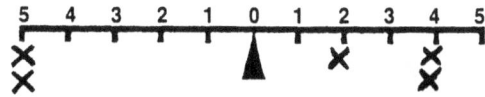

$$5 \times 2 = 10 \qquad 2 + 4 \times 2 = 10$$

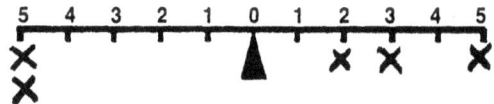

$$5 \times 2 = 10 \qquad 2 + 3 + 5 = 10$$

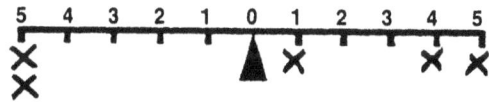

$$5 \times 2 = 10 \qquad 1 + 4 + 5 = 10$$

2a. 2b. 2c. 2d.

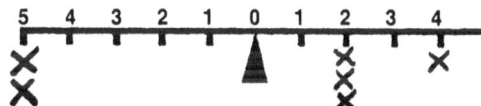

$$5 \times 2 = 10 \qquad 2 \times 3 + 4 = 10$$

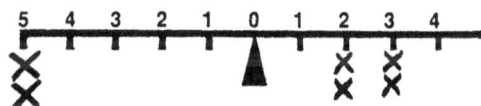

$$5 \times 2 = 10 \qquad 2 \times 2 + 3 \times 2 = 10$$

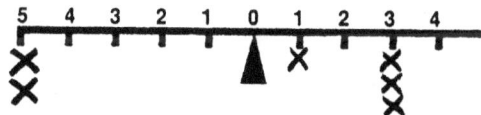

$$5 \times 2 = 10 \qquad 1 + 3 \times 3 = 10$$

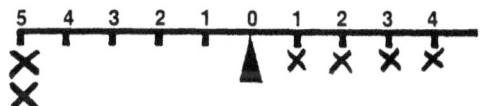

$$5 \times 2 = 10 \qquad 1 + 2 + 3 + 4 = 10$$

MORE BALANCE PUZZLES

Solve each balance puzzle and write the correct equations. *First put clips on the left side as shown...*

...then follow the instructions on the right to balance!

No tab may have more than 1 clip.

a.

$$5 \quad 4 \quad 3 \quad 2 \quad 1 \quad 0 \quad 1 \quad 2 \quad 3 \quad 4 \quad 5$$

$$4 \times 3 + 3 = 15$$

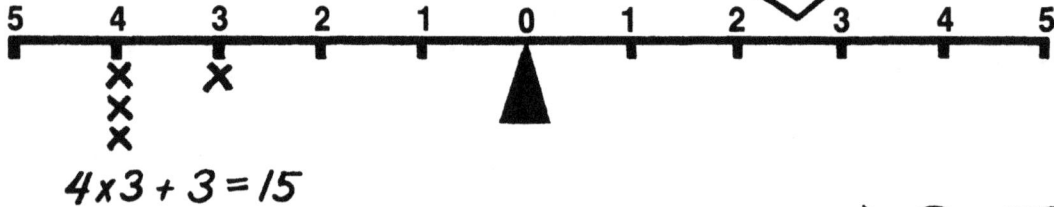

Add clips to only 1 tab.

EQUATION!

b.

$$5 \quad 4 \quad 3 \quad 2 \quad 1 \quad 0 \quad 1 \quad 2 \quad 3 \quad 4 \quad 5$$

Add 2 clips to 2 different tabs.

c.

$$5 \quad 4 \quad 3 \quad 2 \quad 1 \quad 0 \quad 1 \quad 2 \quad 3 \quad 4$$

DON'T USE #5

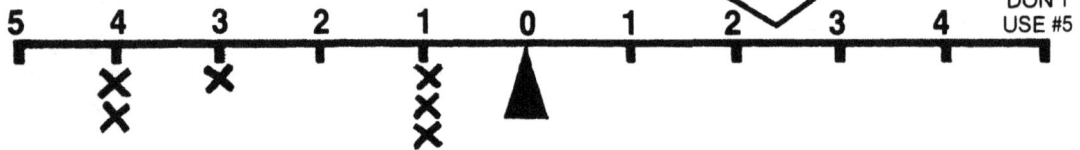

Each tab must have the same number of clips.

d.

$$5 \quad 4 \quad 3 \quad 2 \quad 1 \quad 0 \quad 1 \quad 2 \quad 3 \quad 4$$

DON'T USE #5

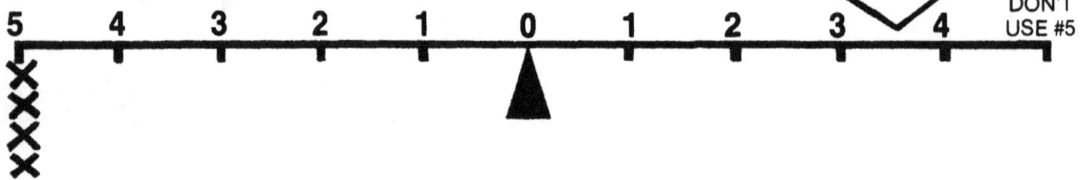

Objective

To practice expressing complex balance combinations as mathematical equations.

Introduction

None required.

Lesson Notes

Some students will likely complete this activity without using a math balance. Others may need to use the balance in order to solve the puzzles. Either way, this exercise establishes a strong connection between mathematics and the nature of balancing beams.

When writing equations, students generally write down the *position* number first, followed by the *number of clips* at that position. (See the example equation, in problem a.) Alternate forms, of course, are also acceptable, as long as the equations are mathematically correct.

Check Point

a.

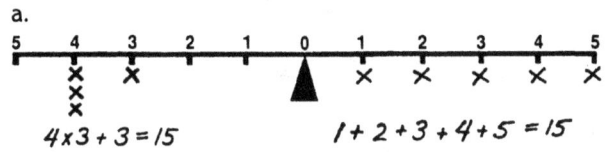

$4 \times 3 + 3 = 15$ $1 + 2 + 3 + 4 + 5 = 15$

b.

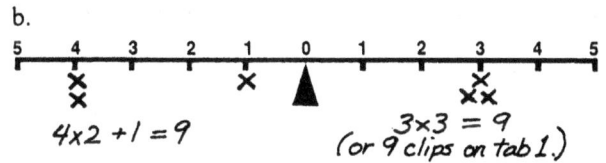

$4 \times 2 + 1 = 9$ $3 \times 3 = 9$
(or 9 clips on tab 1.)

c.

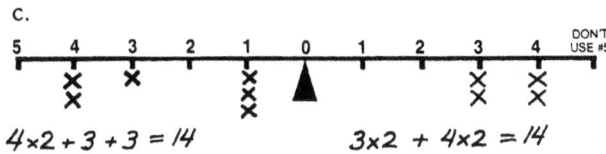

$4 \times 2 + 3 + 3 = 14$ $3 \times 2 + 4 \times 2 = 14$

d.

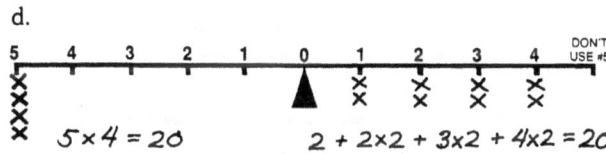

$5 \times 4 = 20$ $2 + 2 \times 2 + 3 \times 2 + 4 \times 2 = 20$

DOES IT BALANCE?

Use math to predict if each beam will balance. Then test each prediction on your balance.

Use math to PREDICT...

...then TEST your prediction.

a. EXAMPLE

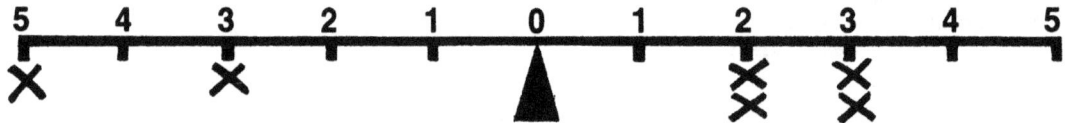

$$5 + 3 = 8$$

$$2 \times 2 + 3 \times 2 = 10$$

PREDICTION: *It won't balance.*

RESULT: *I was right.*

b.

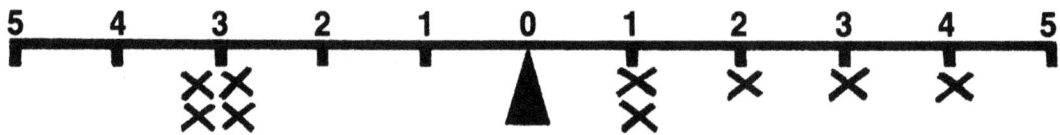

PREDICTION: _____ RESULT: _____

c.

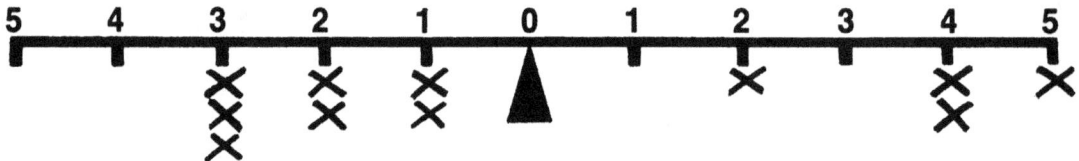

PREDICTION: _____ RESULT: _____

d.

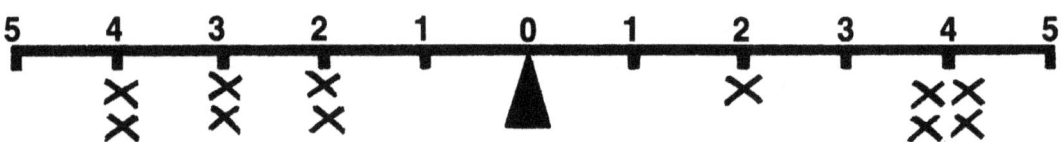

PREDICTION: _____ RESULT: _____

e.

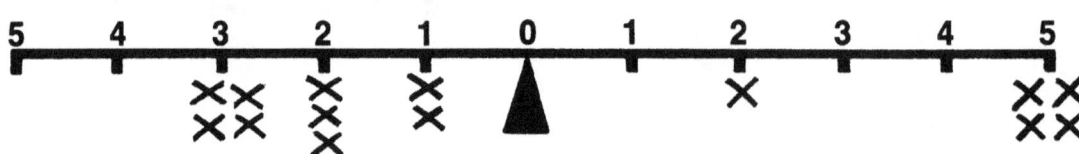

PREDICTION: _____ RESULT: _____

Objective

To mathematically predict and then verify a state of balance or imbalance on a math beam.

Introduction

None required.

Lesson Notes

This activity requires students to predict a state of balance (or imbalance) based on their mathematical calculations, then verify by observation that this prediction is correct. Thoughtful prediction followed by experimental verification is fundamental to the process of doing science. Insist, therefore, that your students do it right. They must follow these steps *in sequence*: (1) calculate, (2) predict, (3) verify.

Check Point

b. Left: 4 x 3 = 12
   Right: 2 + 2 + 3 + 4 = 11
   Prediction: Won't balance. Result: Correct.

c. Left: 3 x 3 + 2 x 2 + 2 = 15
   Right: 2 + 4 x 2 + 5 = 15
   Prediction: Will balance. Result: Correct.

d. Left: 4 x 2 + 3 x 2 + 2 x 2 = 18
   Right: 2 + 4 x 4 = 18
   Prediction: Will balance. Result: Correct.

e. Left: 3 x 4 + 2 x 3 + 2 = 20
   Right: 2 + 5 x 4 = 22
   Prediction: Won't balance. Result: Correct.

WHICH WAY?

Use math to predict
if each balance will. . .
    tilt right,
    tilt left,
    or balance.
Then test your prediction.

PREDICT first...   *...TEST second.*

a.

EXAMPLE

$4 + 3 + 2\times2 = 11$ $2 + 4\times2 = 10$

PREDICTION: *It will tilt LEFT.*   RESULT: *I was right.*

b.

PREDICTION: RESULT:

c.

PREDICTION: RESULT:

d.

PREDICTION: RESULT:

e.

PREDICTION: RESULT:

Objective

To mathematically predict and then verify whether a balance beam tilts left, right, or remains level.

Introduction

None required.

Lesson Notes

Students must again follow these steps in sequence as they solve each problem: (1) calculate, (2) predict, (3) verify.

Extension

In this classroom demonstration, one student's weight is calculated by him or her balancing on a wooden beam opposite another student of known weight. This "teeter-totter" system is completely analogous to a paper clip math balance: the numbered positions on the paper beam correspond to the distance that each student stands from the center pivot. The paper clips used on the paper beam correspond to the weight in pounds (or kilograms) of each student standing on the wooden beam.

PREPARATION

Mark a wooden beam of approximately 3 meters in length (an 8-to-10 foot two-by-four) into 16 equal subdivisions. This is best accomplished by cutting a piece of string equal to the length of the beam, then folding it into half lengths, quarter lengths, etc.

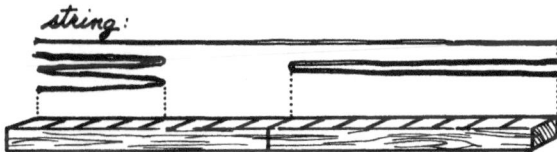

Use a block of wood or a brick as the pivot. If the beam will not balance at your center mark, add a flat rock or other appropriate weight as a "rider."

Obtain a bathroom scale in pounds or kilograms to weigh students and check experimental results.

PRESENTATION

Ask for two volunteers from your class. Weigh one of them on the bathroom scale. Calculate the weight of the other using balance beam mathematics. CAUTION: the wooden beam tends to be unstable when stood upon. Use "spotters" to help your volunteers balance safely.

$$\text{known weight} \times \text{distance from pivot} = \text{unknown weight} \times \text{distance from pivot}$$

Check Point

b. Left: $5 \times 4 + 3 \times 3 = 29$
   Right: $1 + 2 + 4 \times 6 = 27$
   Prediction: Tilts left. Result: Correct.

c. Left: $4 \times 3 + 2 = 14$
   Right: $2 \times 6 + 5 = 17$
   Prediction: Tilts right. Result: Correct.

d. Left: $4 + 5 + 3 \times 2 + 2 + 1 = 18$
   Right: $2 \times 5 + 4 \times 2 = 18$
   Prediction: Will balance. Result: Correct.

e. Left: $3 \times 3 + 2 \times 3 + 3 = 18$
   Right: $4 \times 2 + 5 \times 2 = 18$
   Prediction: Will balance. Result: Correct.

D. COMPARING MASSES

Do you suffer from a shortage of science equipment in your classroom? Do too many students crowd around too few mass balances so that most watch while only a few actually experiment? Not any more.

Like the math balance before, our TOPS equal-arm balance folds from a single sheet of paper. Thread and paper clips support the weighing pans. No shortage of science equipment here. Every student can build one.

Students first use their balances to compare seed masses. Then they weigh things in paper clips. Finally they use real grams.

TOPS gram masses are really paper weights. A 2-gram mass simply uses twice as much paper as a 1-gram mass. These masses are true and the balance is sensitive — to two-hundredths of a gram. It's an impressive scientific instrument, and it's all improvised from simple things.

EVALUATION

Each question evaluates a single activity from COMPARING MASSES as numbered. Use any combination to frame a formal exam or an informal review: Copy these questions on your blackboard, construct your own ditto master, or photocopy the questions while masking out the rest of the page. Evaluate in ways that suit your own teaching style, enabling your students to learn and enjoy science.

Questions

D-3
Ten coins weigh 52 paper clips, and 10 paper clips weigh 150 staples. How many staples does 1 coin weigh?

D-4
Find the weight of each piece of paper in *whole* paper clips.

D-5
Find the weight of each piece of paper to the nearest *tenth* of a paper clip.

D-6
A sealed jar is full of lemon drops. How could you find the number of candies inside this jar without breaking the seal?

D-7
You are given a set of 12 gram masses:

10g, 5g, 2g, 1g, 1g, .5g,

.2g, .1g, .1g, .04g, .04g, .02g.

How would you combine these masses to equal the following:

a. 8.00 g   c. 12.62 g
b. 3.40 g   d. 17.58 g

D-8
5 bottle caps weigh 6 grams;
20 bottle caps weigh 24 grams.

a. Draw a pair of coordinates like this and graph the data above.

b. Show how to find the mass of 15 bottle caps on your graph.

D-9
A chocolate bar has a mass of 28.2 grams.
How much chocolate is this expressed in paper clips?

(1 paper clip = 4.7 grams)

D-10
A shelled peanut has an average mass of .25 grams; an unshelled peanut .30 grams.

Complete the data table. Then graph how the mass of each kind of seed increases with number. Label each graph line.

# of SEEDS	0	5	10	15	20
SHELLED					
UNSHELLED					

Answers

D-3
Using unit analysis,

$$\frac{150 \text{ staples}}{10 \text{ paper clips}} \times \frac{52 \text{ paper clips}}{10 \text{ coins}} =$$

$$\frac{15 \text{ staples}}{\text{paper clip}} \times \frac{5.2 \text{ paper clips}}{\text{coin}} =$$

$$\frac{78 \text{ staples}}{\text{coin}}$$

D-4
(Prepare by cutting and folding 3 pieces of paper to weigh a different numbers of *whole* paper clips. Label them A, B, C).

D-5
(Prepare by cutting 3 pieces of paper to random sizes. Fold and label them X, Y, and Z, then weigh each to the nearest *tenth* of a paper clip. Students answers should not differ by more than a tenth of a paper clip from your "official" values.)

D-6
Balance the sealed jar against an identical empty jar on an equal-arm balance. Add lemon drops to this empty jar, counting as you go, until it counterbalances the full jar.

D-7
a. 8.00g = 5g + 2g + 1g
b. 3.40g = 2g + 1g + .2g + .1g + .1g
c. 12.62g = 10g + 2g + .5g + .1g + .02g
d. 17.58g =
   10g + 5g + 2g + .5g + .04g + .04g

D-8
a.

b. Fifteen bottle caps on the graph is associated with a mass of 18 grams.

D-9
28.2 g x 1 p.c./4.7 g = 6 p.c.

D-10

# of SEEDS	0	5	10	15	20
SHELLED	0	1.3	2.5	3.9	5.0
UNSHELLED	0	1.5	3.0	4.5	6.0

SEQUENCING

COMPARING MASSES is a prerequisite to E because the gram balance developed here is used in several E activities as well. If your class has not yet completed C, refer back to C-2 for directions on assembling the base to this balance.

Related Activities: C---**D**—E---F---G

MATERIALS

Here is everything your students will use for the next 10 activities on COMPARING MASSES. Materials printed in normal type are part of the core 15-things-in-a-box inventory that support all 100 activities. Materials printed in *italics* are additional local materials that you provide or ask your students to bring from home. Pencil and paper are already assumed and therefore unlisted. Each item is numbered with the activity where it is first used.

(D-1) Scissors.
(D-1) Clear tape.
(D-1) Straight pins.
(D-1) Clothespins with masking tape"ears" attached to cans or bottles. See steps 1-3 in activity C-2.
(D-2) Thread.
(D-2) Paper clips of uniform size and weight.
(D-3) *Four kinds of seeds.* We recommend using *beans* (pinto beans), *corn* (popcorn), *lentils* and *rice* (white long-grained variety). If some of these seeds are not available in your area, substitute other seeds that have the same relative size. (See "Preparation" in teaching notes D-3.)
(D-4) *Objects to weigh.* Collect at least 8 small objects including a *coin, sheet of paper, bottle cap* and *pen cap.* Anything that fits into the pan is appropriate to weigh. Other suitable objects might include a nail, a short pencil, an eraser, a piece of chalk, a cork or a button.
(D-4) *Scratch paper.*
(D-5) *Lined notebook paper.*

FURTHER STUDY

Use problems like these plus "extension" ideas in COMPARING MASSES to lead your students beyond worksheet activity into original research and investigation. Each discovery leads to more questions, deeper questions, better questions than these. Answering them is what good science is all about.

Imagine that you are living long, long ago. Nobody in your culture has ever dreamed about weighing anything before. Write a story telling how you invented a balance beam for the first time.

What is the difference between the *mass* of an object and its *weight*? Research this important distinction and write a report.

Read about other kinds of instruments that balance and weigh:
    torsion balance
    spring scale
    steelyard balance

Can you improvise any of these other systems using simple materials?

BUILD A PAPER BEAM BALANCE (1)

1 Carefully trim this paper along the dotted lines.

CUTOUT D-1

2 Fold it in half exactly along the center line.

Keep the printing on the outside.

3 Fold up this half again so that all the edges meet.

FOLD UP BOTH EDGES TOGETHER.

4 Fold up this quarter again so the edges meet.

This is your third fold.

5 Tape where shown to form a single closed strip.

6 Poke a pin through each crossmark. Enlarge both end holes (not the middle) by wiggling the pin.

7 Write your name here.

R

CENTER

FOLD

L

NAME:

Objective

To fold a paper beam to use in an equal-arm balance.

Introduction

This worksheet folds into an actual paper beam balance, just like the math balance previously constructed. The directions are similar as well, but less complicated. Students who have already built the math beam should easily be able to complete this balance beam as well, with little outside help. The only introduction you need provide is to construct this balance in advance, following directions in D-1 and D-2. Display this balance as a model for your students to follow, then stand clear and let construction begin.

Lesson Notes

2. When folded together along the middle guideline, the top and bottom edges of the paper don't quite meet. These edges will "creep" together as the beam is doubled over in step 3 and again in step 4.

Some students may fold the paper inward and thereby cover the directions. The illustration directs students to fold the worksheet outward.

3. *Both* bottom edges fold up as illustrated, not just one. Students who fold up just the first edge by mistake, tend to wait until step 4 before folding up the second edge. Missing the 3rd fold entirely, they end up in step 5 with a beam that is twice as wide as it should be.

6. To insure that the balance beam has arms of equal length, the pin must penetrate the exact point of intersection inside each circle. You may need to supervise younger children to help them accurately place each pin point.

You'll need to apply considerable pressure with your finger each time you force the pin through all the layers of paper in the beam. Wrapping your finger in masking tape protects the skin from being pushed in excessively by the pinhead.

Leave the *middle* pinhole small, but enlarge both *end* pinholes. Work the pin back and forth making these as large as possible. If these end holes remain too small, it will be difficult later on to force the blunt end of a paper clip through each hole. (See step 6 in Activity D-2.)

7. The completed equal-arm beam should look like this:

Check Point

Is the paper well folded so the edges match more or less evenly? Are the three circled cross-marks punched through right on center? Are the two end holes enlarged sufficiently to receive a paper clip? Is the middle pin hole left small? (If it was enlarged by mistake, cover it on both sides with small patches of masking tape, then repunch the hole.)

BUILD A PAPER BEAM BALANCE (2)

1 Cut out both large squares. Fold along the diagonals.

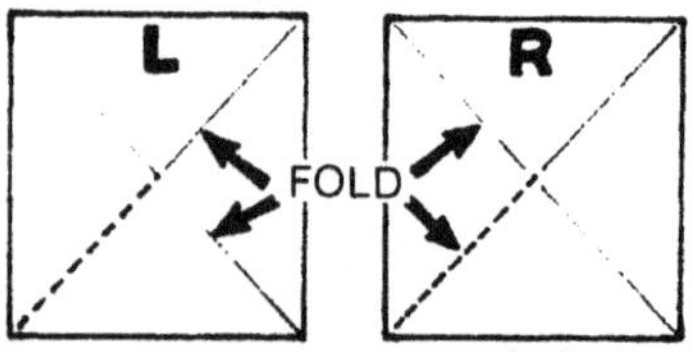

L   R   FOLD

2 Cut on the diagonal line to the center of each square.

L   R

3 Fold each square into a pyramid.

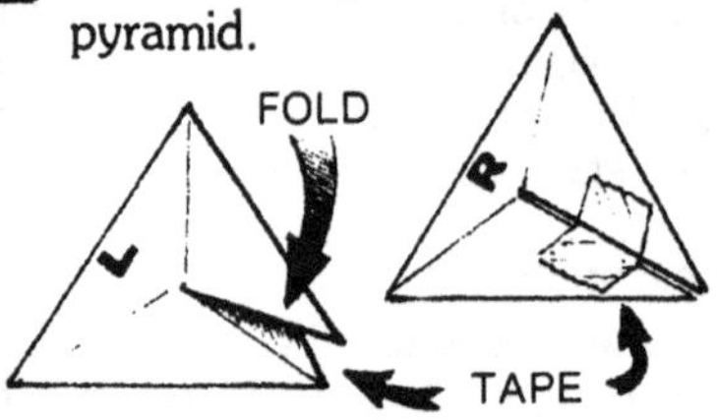

FOLD   R   L   TAPE

4 Cut a thread as long as your arm,

then triple it.

Tie one end around a pencil.

Trim both ends.

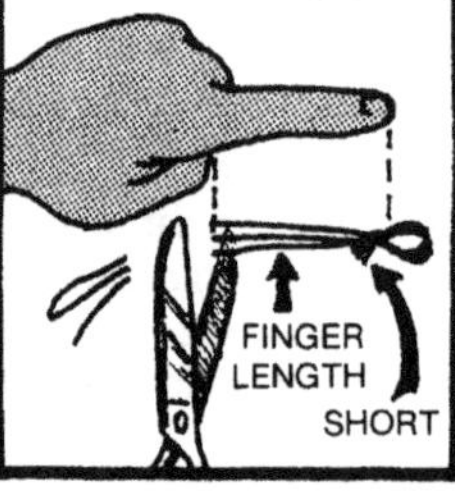

FINGER LENGTH   SHORT

Make another the same length.

5 Tape the threads to each pyramid, forming 2 baskets.

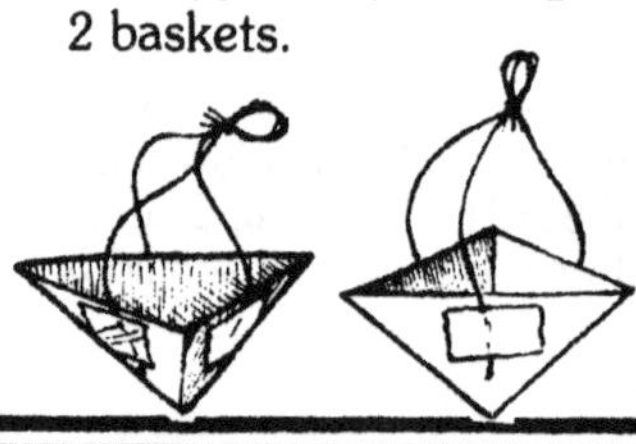

6 Bend out 2 paper clips like this. . .

...then push each one into an end of the balance.

MAKE A HOOK

7 Set up your balance. Hang the correct basket at each end.

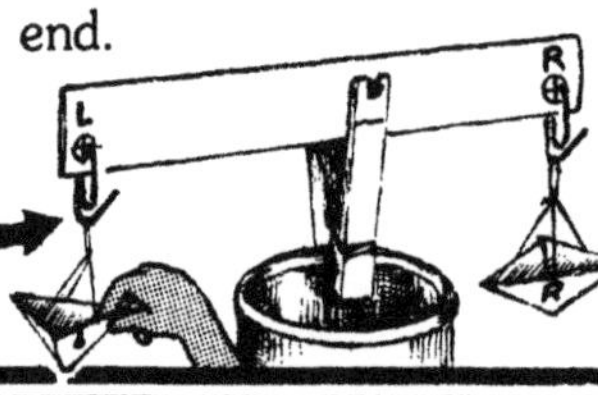

L   R

Add a tape rider to balance level.

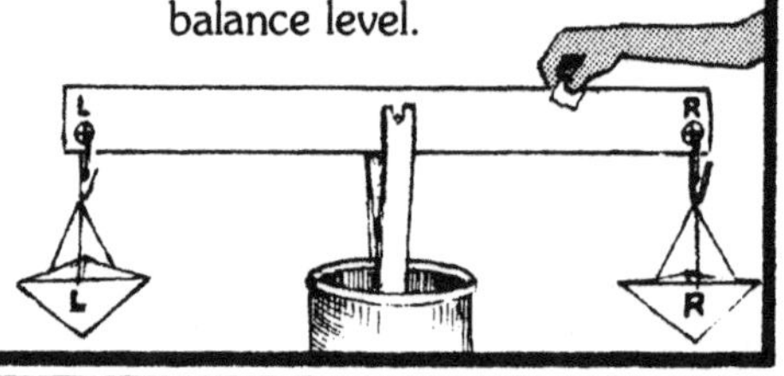

L   R

CUTOUTS D-2

L   R

Objective

To complete construction of an equal-arm balance.

Introduction

None required.

Lesson Notes

1-2. Notice that the diagonals are first folded in step 1, before the dashed lines are cut in step 2.

4. The threads should be cut to about finger length — neither longer, nor shorter. When cut too long, the weighing pans tend to drag on the table surface. When cut too short, there is not enough clearance to easily transfer objects in and out of the pans.

5. Adjust these threads so the baskets hang at about the same height. This is for cosmetic purposes only. The balance functions just as well if the baskets hang at unequal heights, but it looks rather odd.

6. These end pinholes have *already* been enlarged by working a pin back and forth (see step 6 in activity C-1). This step continues this enlargement process using a heavier drill. Little by little the hole will open until you can force the paper clip through.

You can also use a sharp pencil. Place the pencil point on the pin hole, then rotate it back and forth between your fingers. This will produce a clean, well-defined hole.

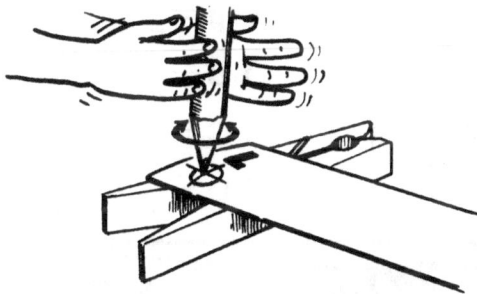

The pencil must be good and sharp. Using a dull pencil will only dog-ear the corners of the beam. Be careful not to drill the holes too large. Open them no bigger than a pinhead. This allows an inserted paper clip to rotate freely, without wobbling from side to side.

Notice that the paper clip is inserted past the first curve in the wire. Continue to push the wire through the hole until you rotate the paper clip "hook" into an "up" position.

7. Students should now reinsert a pin through the center hole in the beam, then rest it on the same clothespin support system that they built for their math balance. If your class has not yet constructed this support system, direct students to study the directions in activity C-2, steps 1-3.

Even though the left and right pans have the same size, it is likely that one side of the balance beam will still be lighter or heavier than the other, due to differences in the amount of tape and thread used to construct each side. A tape rider, of appropriate size, compensates for this weight difference. Placing it somewhere on the higher (lighter) side of the beam, brings it back to a level (centered) position. (Students who have not previously made riders should consult activity C-2, step 7.)

Because each pan has a slightly different mass, students cannot interchange them and expect the beam to remain centered. To avoid repositioning the rider more than necessary, the right pan should always be attached to the right side of the beam (both marked "R"), and the left attached to the left (both marked "L").

Notice that the beam will not balance upside-down. This places the pivot pin too low relative to the center of mass of the beam, thus creating instability.

Check Point

Does the balance return to a centered level position each time the beam is tipped off center? Do the paper clips rotate freely at each end when the beam is tilted from side to side?

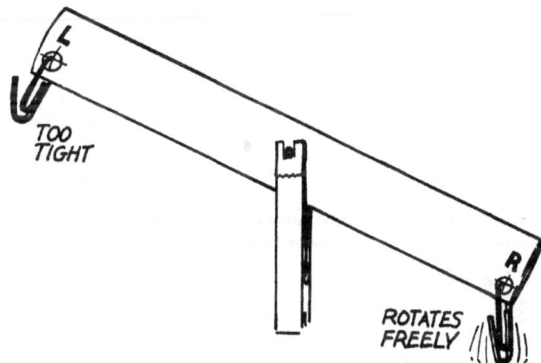

SEEDS AND PAPER CLIPS

1   Write the names of seeds, from largest to smallest, under each letter.

a.	b.	c.	d.	e.
paper clip				

2   Center your empty balance.

Move the rider if you need to...

a. Solve each equation using your balance.

10 paper clips = ... B

10 B = ... C

10 C = ... D

10 D = ... E

b. Divide each equation by 10.

Move your decimal points.

1 paper clip = ... B

1 B = ... C

1 C = ... D

1 D = ... E

3   Use your results above to mathematically solve each equation below. . .

Hint: You need to multiply.

a.	b.	c.
1 paper clip = ? C	1 B = ? D	1 C = ? E
_____	_____	_____

...then check each answer on your balance. Write your results here.

Objective

To make mass comparisons on a balance beam and thereby generate simple mathematical relationships.

Preparation

This activity uses four kinds of seeds: beans (pinto beans), corn (popcorn), lentils and rice (white, uncooked, long-grained). If some of these seeds are not available in your area, substitute local varieties. As you gather seeds, keep 2 things in mind: (1) The seeds within each particular variety should be roughly *uniform* in size. (2) The size of all 4 varieties (plus paper clips) should be distinctly *different* from each other. (If paper clips are no longer your heaviest group, change their relative position on the worksheets from "A" to another letter.)

If you made seed substitutions, try working through this experiment yourself before your students try it. Write down your answers and develop a custom answer key, based on the seeds you use.

Introduction

To prepare your class for the math they will encounter in step 3, hold a class discussion about money. Write equations on your blackboard that relate 4 familiar monetary units in descending value. In Canada or the United States, for example, you might relate dollars, quarters, nickels and pennies:

$$1 \text{ dollar} = 4 \text{ quarters}$$
$$1 \text{ quarter} = 5 \text{ nickels}$$
$$1 \text{ nickel} = 5 \text{ pennies}$$

Given these equations, discuss how you would solve these related equations:

$$1 \text{ dollar} = ? \text{ nickels}$$
$$1 \text{ quarter} = ? \text{ pennies}$$

One powerful way to solve problems is to keep track of units. Equations can also be expressed as fractions. Fractions can be multiplied in any order, or turned upside down, so that units of known quantities cancel out to yield the units you're looking for:

$$\frac{? \text{ nickels}}{1 \text{ dollar}} = \frac{4 \text{ quarters}}{1 \text{ dollar}} \times \frac{5 \text{ nickels}}{1 \text{ quarter}}$$

$$\frac{? \text{ pennies}}{1 \text{ quarter}} = \frac{5 \text{ nickels}}{1 \text{ quarter}} \times \frac{5 \text{ pennies}}{1 \text{ nickel}}$$

$$\frac{? \text{ pennies}}{1 \text{ dollar}} = \frac{4 \text{ quarters}}{1 \text{ dollar}} \times \frac{5 \text{ nickels}}{1 \text{ quarter}} \times \frac{5 \text{ pennies}}{1 \text{ nickel}}$$

Younger students might simply be instructed to multiply:

$$1 \text{ dollar} = ? \text{ nickels}$$
$$4 \times 5$$
$$1 \text{ quarter} = ? \text{ pennies}$$
$$5 \times 5$$

Lesson Notes

2a. It is likely that a specific number of seeds may not quite balance the beam, then adding one more seed will tip the balance too far. Don't worry. Answers in this step are approximate at best.

2b. To fill in this second series of equations, students simply need to divide each equation by 10 (move the decimal 1 place to the left). No experimentation is required here, though some students may attempt to use their balances.

3. When checking their math, students should select average-size seeds to weigh and compare. Agreement will only be approximate, due to variations in seed size.

Check Point

(Because there is considerable size variation among seeds of the same kind, particularly among pinto beans and popcorn seeds, these answers are only approximate. If you substitute different seeds, of course, you will need to develop a different answer key.)

1. a. paper clips b. pinto beans c. popcorn d. lentils e. rice

2a. 10 paper clips = 12 pinto beans
10 pinto beans = 38 popcorns
10 popcorns = 24 lentils
10 lentils = 31 rice grains

2b. 1 paper clip = 1.2 pinto beans
1 pinto bean = 3.8 popcorns
1 popcorn = 2.4 lentils
1 lentil = 3.1 rice grains

3a. 1.2 x 3.8 = 4.6 popcorns
(5 popcorns on the balance)

3b. 3.8 x 2.4 = 9.1 lentils
(10 lentils on the balance)

3c. 2.4 x 3.1 = 7.4 rice grains
(8 rice grains on the balance)

PAPER CLIP MASSES (1)

1 Find the mass of at least 8 small objects using paper clips. Use items in this list, plus some of your own.

10⁻ means it weighs a little LESS than 10.

10⁺ means it weighs a little MORE than 10.

a coin	p.c. ◄— (paper clip)
a folded paper	
a bottle cap	
a pen cap	

Write your initials on all objects.
Save them in your can to use again.

Use 4 items of your own here.

2 Find the mass of just one rice grain in paper clips. Explain how you did this.

What part of me makes one of you?

3 Is it better to measure mass in paper clips or rice grains? Explain why you think so.

Objective

To find the mass of common classroom objects using a paper-clip standard of measure.

Introduction

Any number of conditions or events — moisture, accumulated dirt, interchanging the balance pans, rough handling — can all conspire to tilt a beam off center. Emphasize the importance of checking often to insure that the empty balance still centers level. Measuring with an uncentered balance is like using a ruler the doesn't start at zero. The bad results are simply not worth the effort.

Lesson Notes

1. Few objects weigh an exact number of whole paper clips. Students should write a *plus* after their answer for items that weigh a little more than a whole number of clips; a *minus* for objects that weigh a little less.

Be sure students identify all 8 objects with their initials, or some other mark. They'll need to weigh these same objects again in activity D-5 (to the nearest tenth of a paper clip) and activity D-9 (to the nearest tenth of a gram).

In a right-handed world it is customary to put the object to be weighed in the left *pan* and then add units of mass (in this case, paper clips) to the *right* pan. Whether your students follow this convention or not, they should consistently use one pan exclusively for holding objects to be weighed, the other for holding mass units.

Why is such consistency important? All equal-arm balances, to some extent, are unequal (biased). If one arm of the balance, is slightly longer than the other, it measures slightly heavier when holding the object to be weighed, slightly lighter when holding the paper clips (or other mass unit). Data that is *always* a little too high or too low, because of consistent pan use, is more reliable than data that is one time too high, the next time too low, because of inconsistent pan use.

Balance bias is something to be aware of, but not worry about. Balances that were properly constructed are *already* accurate enough. Perfectionists keen on achieving even greater accuracy should see the introduction to D-5.

Check Point

(Answers may vary, depending on the mass of paper clips used and the kinds of objects weighed.)

1. coin (penny) = 5^+ p.c.
   folded paper (standard notebook paper) = 7^- p.c.
   bottle cap = 4^- p.c.
   pen cap = 1^+ p.c.
   (Students should list 4 additional items here.)

2. There is considerable size variation among rice grains. Here is one possible answer:
   30 rice grains = 1 p.c
   Thus, 1 rice grain = 1/30 p.c.
   = .033 p.c.

3. Paper clips are a better standard of mass because they are more uniform in size. Rice grains, by contrast, vary in size; many are chipped and broken. If, however, you wish to find the mass of a very light object, then lighter rice grain units are a more appropriate (but less uniform) standard.

NAME: CLASS:

PAPER CLIP MASSES (2)

1 Carefully cut out the strip of lined paper.

2 Cut parallel strips off the end until it has a mass of exactly one paper clip.

ONE PAPER CLIP

Cut off **THIN** strips.

3 Divide this "paper clip strip" into 10 equal parts.

1 2 3 4 5 6 7 8 9 10

LINED PAPER will help you do this.

4 Cut your strip to make these paper clip fractions:

| 1 | 2 | 3 | 4 | 5 |
.5 pc

6 7
.2 pc

8 9
.2 pc

10
.1 pc

LABEL EACH PART.

5 **a.** Find each mass as before, this time to the nearest *tenth* of a paper clip.

a coin	
a folded paper	
a bottle cap	
a pen cap	

b. Compare these mass values to what you found before.

SAVE YOUR OBJECTS

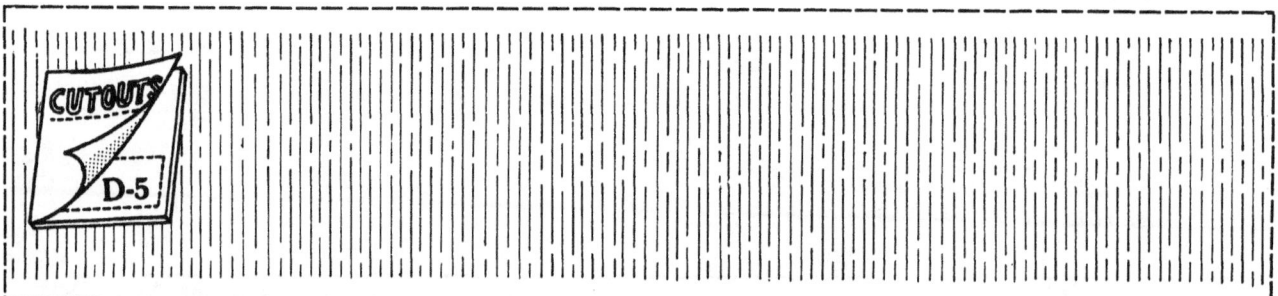

CUTOUTS
D-5

Objective

To find the mass of common classroom objects using a mass standard refined to a tenth of a paper clip.

Introduction

Equal-arm balances don't have arms that are exactly equal. Because of measuring error (which can be minimized but never entirely eliminated) one side of any balance, measured from pivot pin to paper clip, will be slightly longer or shorter than the other side. If this difference is large (more than a few millimeters) the balance is *biased*. Objects will appear to weigh heavier or lighter depending on the weighing pan you put them in. If this difference is small (less than 1 millimeter) the balance is nearly *true*. Objects will appear to weigh the same, no matter which balance pan you use.

If your students have carefully constructed their equal-arm balances, they should be true enough to perform well in all TOPS experiments requiring mass measurements. Greater standards of accuracy are not necessary, yet they are possible to achieve. If you want your students to fine-tune their balances, and understand the physics of balancing in greater depth, then read on. Otherwise, skip over this introduction entirely.

Here is a way to reduce balance bias to an absolute minimum: First ask each student to center his or her empty balance. Move the rider, as necessary, to make the beam center dead level each time it is tipped off balance. Now ask students to add 5 identical paper clips to each pan, and observe the direction of tilt (if any). Switch paper clips to opposite pans and observe the tilt again (if any):

● The balance has unequal arms if the beam tilts the *same* way both times. Adjust it by moving the center pivot pin just a tiny bit in the direction of the arm that points down (the longer arm.)

● The balance is nearly true (has equal arms) if the beam remains level both times.

● The balance is also nearly true if the beam tilts first one way, then the opposite way by the same amount. In this case the paper clips have slightly different masses.

Lesson Notes

2. The idea here is to cut off slices of paper until the strip is reduced to a mass of 1 paper clip. Students must start with a centered balance before they begin this step.

As they approach a mass of 1 paper clip, caution students to shave off smaller and smaller slices of paper. Those who fail to exercise enough patience may cut off more than they intend. If the total weight of this strip falls significantly below 1 gram, the only recourse is to start over, using an equivalent piece of plain paper cut to size. Students must *not* add tape to remedy any mass deficit. This strip can only be subdivided into accurate fractions of a gram (in step 4) if its mass is *evenly* distributed along its entire length.

3. Students often confuse the number of lines they must draw (9) with the number of spaces they need (10). Numbering the *spaces* , as shown, reduces this confusion.

4-5. These fractional gram masses, used in combination with whole paper clips, can now be used to weigh objects with greater accuracy. First add as many whole paper clips as possible, without exceeding the mass. Then continuing adding smaller paper masses to weigh the object to the nearest *tenth* of a clip.

Check Point

(Answers may vary, depending on the mass of paper clips used and the kinds of objects weighed.)

5a. coin (penny) = 5.2 p.c.
folded paper (standard notebook paper) = 6.8 p.c.
bottle cap = 3.9 p.c.
pen cap = 1.3 p.c.

(Students should weigh 4 additional items, the same ones they used in activity D-4.)

5b. These mass values agree with the values in activity D-4, except they are accurate to the nearest tenth of a paper clip instead of the nearest whole paper clip.

NAME: CLASS:

EDUCATED GUESS

1 Start with 2 sheets of paper that are exactly the same size and weight.

2 Ask a friend to hide from 1 to 10 paper clips in one of the papers while you look away.

3 Use your balance to guess how many paper clips are wrapped inside.

a. If you guessed wrong, try again.

b. If you guessed right, tell how you did it.

4 As a bank teller your job is to count coins to wrap in rolls of 50.

Tell how you would use a balance to make your job easier.

Copyright © 1988 by TOPS Learning Systems. Reproduction limited to personal classroom use only.

Objective

To count an unknown quantity of paper clips by comparing their mass to a known number of paper clips on a balance.

Introduction

None required.

Lesson Notes

3. Some students may forget to put the second sheet of paper in the other pan to counter-balance the paper that wraps the clips. They will erroneously predict a number of clips that is too large and need to repeat the experiment, as directed. In time, they'll discover and fix their own error: independent problem solving at its best.

After adding this extra paper, students should next add paper clips one at a time to the same pan until the beam balances. Other may follow a less efficient "hit or miss" strategy, wrapping up the number of clips they think is correct, then adding them all at once to the balance. Either method leads to a correct guess.

This hide-and-guess game is very popular with younger students. Some may wish to repeat the procedure several times beyond their first correct answer. This is fine.

4. An easier way to count coins is to first count out just *one* stack of 50. Then make other piles of coins equal in height to this original stack. This piling method takes advantage of their uniform thickness rather than their uniform mass.

Check Point

3b. Center your balance. Place the wrapped-up clips in one pan and the unused paper (folded up) in the other. Add clips to the lighter (higher) side until your beam balances level. The number of clips you add equals the number wrapped in the paper.

4. Center your balance and place 50 coins in one of the pans. To "count" 50 more coins, simply add them to the other pan until the beam returns to a level position.

NAME: CLASS:

GRAM MASSES

1 Add up the total mass of the GRAM STRIP below.

Add all the decimal fractions.

Total mass:

2 Cut out the *whole* GRAM STRIP.

One big piece— cut carefully!

GRAM STRIP

3 Balance another piece of paper against this GRAM STRIP. Trim until it equals 1 gram (when folded and taped).

L   R

GRAM STRIP   OTHER PAPER

4 Make all these masses using paper, tape, or other materials. Label each one.

1g   2g

1g   10g

1g   5g

Fold each mass TIGHTLY.

5 Carefully cut your GRAM STRIP into its 7 separate parts. Fold so the mass numbers show.

.2g   .04g   .02g

.1g   .04g

.5g   .1g

6 Compare your masses and adjust if needed.

$$2g = 1g + 1g$$
$$5g = 2g + 1g + 1g + 1g$$
$$10g = 5g + 2g + 1g + 1g + 1g$$
$$1g = all\ fractional\ pieces$$

THESE MUST BALANCE!

7 Store all masses in your can.

.5g		.2g	.1g	.1g	.04g
CUTOUTS D-7	GRAM STRIP				.04g
					.02g

Objective

To develop a series of gram masses to use with the equal arm balance.

Introduction

None required.

Lesson Notes

2. If you are photocopying these worksheets instead of using our more convenient and economical *Student Cutouts Booklets*, be aware that the accuracy of this gram strip depends upon the weight of paper you use in the copier. We have sized this gram strip to equal close to1 gram when used with 50 pound bookstock, a standard-weight copy paper.

2-6. The gram masses developed in this activity have remarkable accuracy, as long as each step is completed with care:

In step (2) cut around the gram strip's outside perimeter staying right *on* the line.

In step (3) first make sure the balance is perfectly centered. Then trim another piece of paper until it equals the mass of this gram strip as close as you can get. (If you trim too much, it's OK to add tape.)

In step (4) you'll need to tape the larger paper masses together so they take up less space. Make sure this tape is counted as part of the total mass. (Notice that you can still lighten masses that are taped by trimming the edges.)

In step (5) keep your scissors on the line as you separate gram fractions. (Stray off any line and you'll add extra mass to one fraction while subtracting mass from the other.)

Do all of the above and you'll get masses in step (6) that are internally consistent. That is, many smaller masses in one pan will just about balance an equivalent larger mass in the other.

6. No matter how carefully they make their gram masses, the beam may still tip slightly off center, as students test each of these 4 equalities. Experimental error, after all, is impossible to eliminate completely. If the beam consistently tips in the *same* direction, the downward tilting arm may be too long relative to the upward tilting arm. Check for unequal balance arms using the method outlined in the introduction to D-5.

Check Point

1. $.02 \text{ g} + .04 \text{ g} + .04 \text{ g} + .1 \text{ g} + .1 \text{ g} + .2 \text{ g} + .5 \text{ g} = 1.00 \text{ g}$

PAPER CLIP GRAPH

1 Use your balance and gram masses to complete this table.

CUTOUTS D-8

NUMBER OF PAPER CLIPS:	0	5	10	15	20
MASS in GRAMS					

PLOT YOUR DATA

Circle your points and draw the best straight line you can.

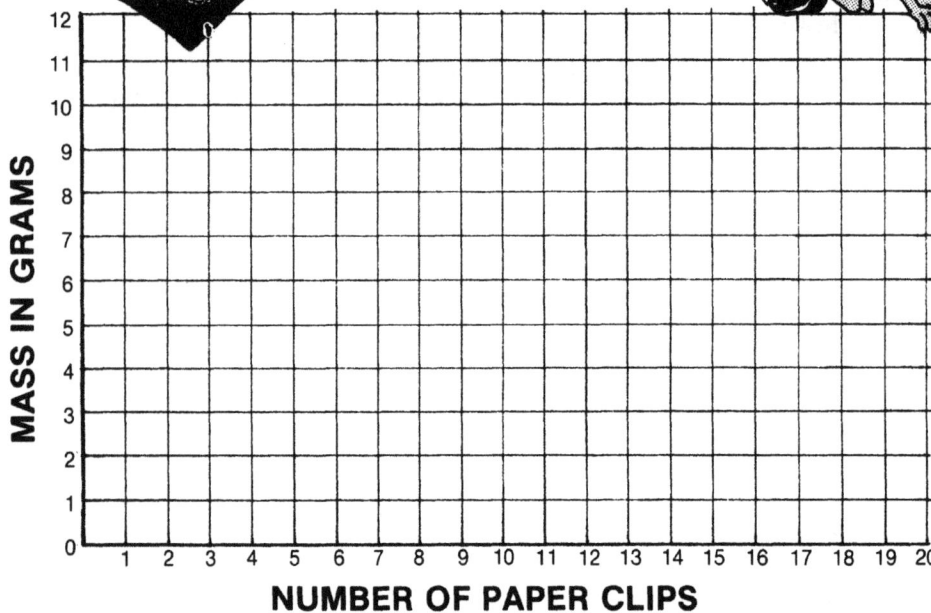

Don't draw through your points.

FOLDED PAPER

A good straight edge!

MASS IN GRAMS — 0 1 2 3 4 5 6 7 8 9 10 11 12

NUMBER OF PAPER CLIPS — 1 2 3 4 5 6 7 8 9 10 11 12 13 14 15 16 17 18 19 20

2 Read your graph to find the mass of one paper clip.

3 Find the mass of 10 paper clips from your data table. Use it to calculate the mass of one paper clip.

4 Which paper clip value is more accurate, the one in step 2 or 3? Explain.

Objective

To graph how the total mass of paper clips increases in direct proportion to their total numbers.

Introduction

Those who are still unfamiliar with graphing may benefit from a short review. Mark off a pair of numbered coordinates on your blackboard, and plot these ordered pairs. Demonstrate how each plotted point on the graph is fixed by perpendicular lines drawn from its two coordinate numbers.

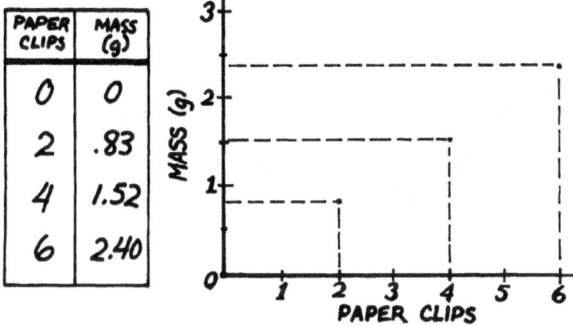

PAPER CLIPS	MASS (g)
0	0
2	.83
4	1.52
6	2.40

Now circle the plotted points and connect them with a smooth line that best represents their overall graph trend, in this case a straight line. Because these points are scattered somewhat by experimental error, some may not meet the line.

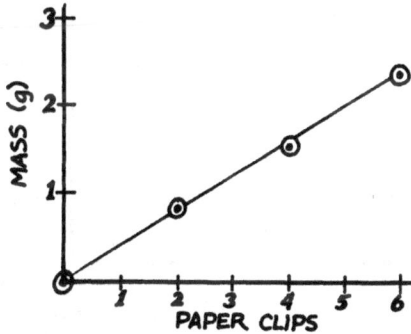

Keeping the graph line outside each circle preserves the clarity of each point inside, allowing others to verify that it has been plotted accurately.

Lesson Notes

1. If you know the mass of 5 paper clips, then 10 paper clips weigh 2 times as much, 15 paper clips 3 times as much, and so on. You can take advantage of this relationship by weighing only the first 5 clips, then completing the rest of the table mathematically. Or you can directly weigh each group of paper clips. Either method produces a good result. Allow your students to follow their own inclinations.

Check Point

Mass numbers will vary, depending on the size of the paper clips used. The overall trend in the graph line, however, must always be a straight line.

1.

NUMBER OF PAPER CLIPS:	0	5	10	15	20
MASS in GRAMS	0	2.35	4.70	7.05	9.40

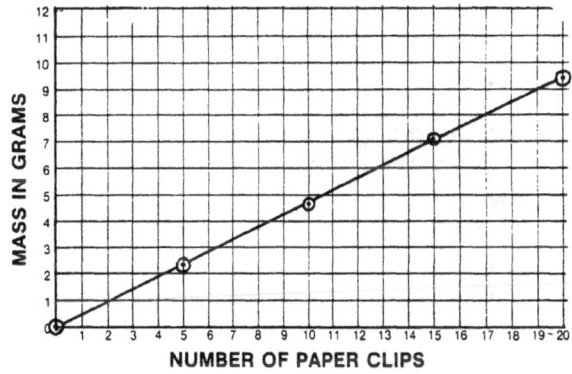

2. One paper clip on the graph above is associated with a mass of approximately .5 grams.

3. Ten paper clips in the above data table have a mass of 4.7 grams. One clip, therefore, must have a mass of .47 grams.

4. The value in step 3 (4.7 grams) is more accurate. It is based on a sampling of 10 clips that were actually weighed. The accuracy of the value in step 2 (.5 grams), by contrast, depends entirely on the placement of the graph line.

PAPER CLIPS TO GRAMS

1 Complete this table.	Use the balance and gram masses you just made.	Use values from Activity D-5.	Use value from Activity D-8.	Multiply.	Yes or No?
OBJECT	**ACTUAL MASS** (grams)	**CALCULATED MASS** $\frac{\text{PAPER CLIPS}}{1} \times \frac{\text{GRAMS}}{\text{PAPER CLIPS}} = $ GRAMS			**GOOD AGREEMENT** ?
a coin					
a folded paper					
a bottle cap					
a pen cap					

2 Which is a better unit of mass, a paper clip or a gram? Explain why.

3 Suppose you want to find the mass of an elephant. Is the gram a convenient unit to use? Explain.

Objective

To convert from paper clip units of mass to standard gram units. To compare calculated mass values with experimental values.

Preparation

This activity requires that students weigh (in grams) the *same* 8 object they weighed before (in paper clips). Make sure they have their *own* objects in hand before beginning. (These were previously identified by personal initial in activity D-4.)

Lesson Notes

1. Students should work *across* this table, not down. This enables them to check the validity of one complete line, comparing actual mass values against calculated values, before proceeding to the next line.

Remind your class that a valid measurement has 2 parts — a number and a unit. Insist that they write both in each box.

THE DIFFERENCE BETWEEN WEIGHT AND MASS

Weight is a measure of gravitational force. It changes as gravity changes. You weigh less on the moon than on Earth, but more on Jupiter. In free space, you weigh nothing at all.

Mass, by contrast, is a measure of the amount of matter an object contains. Your body is made of the same stuff, regardless of where you take it. Your mass is constant on Earth, on the moon, on Jupiter, or in free space.

Popular English make no distinction between these two terms. Mass and weight are inextricably linked by Earth's nearly constant gravitational field. Technically it is incorrect to ask how many *grams* an object *weighs*, even though this sentence makes perfectly good sense. Any English dictionary will tell you that a gram weighs .035 ounces. But this is only true on Earth. A gram weighs .000 ounces in space.

As long as you restrict the use of these TOPS activities to planet Earth, mass and weight don't need to be differentiated. Earth-bound beginning science students may not be intellectually ready to comprehend this difference anyway. If you think your students are ready, then by all means, clarify the distinction.

Extension

To help your class learn to accurately estimate gram masses, hold a contest: Ask who can find a rock that comes closest to a mass of 10 grams. (No broken rocks, please.) Each student should submit their best rock, weighed to the nearest tenth of a gram. You verify the mass of the winning rock.

Check Point

1.

OBJECT	ACTUAL MASS (grams)	PAPER CLIPS 1	×	GRAMS PAPER CLIPS	=	GRAMS	GOOD AGREEMENT ?
a coin	2.48 g	5.2 p.c.		.47		2.44 g	yes
a folded paper	3.22 g	6.8 p.c.		.47		3.20 g	yes
a bottle cap	1.80 g	3.9 p.c.		.47		1.83 g	yes
a pen cap	.58 g	1.3 p.c.		.47		.61 g	yes

2. The gram is a better unit of mass. It is a well-defined, widely-accepted standard of measure recognized and used by scientists around the world. Who besides TOPS students know the mass of paper clips? Paper clips used in Africa can hardly be expected to have the same mass as paper clips used in Asia.

3. The gram is not a convenient unit of measure because it is too small relative to the mass of an elephant. A much larger unit of mass, a metric ton for example, would be more suitable. (1 metric ton = 1,000,000 grams)

SEED GRAPH

1 Write the name of a seed in each box, from LARGEST...

...to SMALLEST.

Use your balance to fill in each table.

CUTOUTS

D-10

a.					**b.**					**c.**					**d.**				
NUMBER OF SEEDS	0				NUMBER OF SEEDS	0				NUMBER OF SEEDS	0				NUMBER OF SEEDS	0			
MASS IN GRAMS	0				MASS IN GRAMS	0				MASS IN GRAMS	0				MASS IN GRAMS	0			

Plot and label 4 graph lines. Finish each one *before* starting the next.

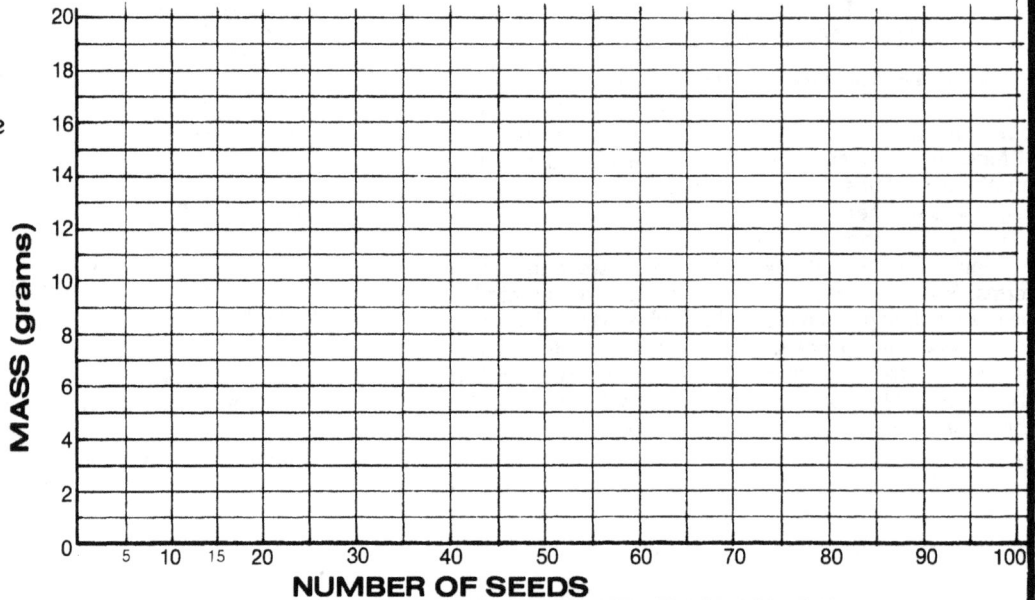

MASS (grams)

20
18
16
14
12
10
8
6
4
2
0

5 10 15 20   30   40   50   60   70   80   90   100

NUMBER OF SEEDS

2 How does the mass of a seed affect the slope (steepness) of its graph line?

3 Predict the graph line for seeds that. . .

a. ...float on air.

b. ...are heavier than coconuts.

Objective

To graph how the mass of seeds increases in direct proportion to their numbers. To understand slope as a function of seed size.

Introduction

None required.

Lesson Notes

1. Weigh and graph the same 4 kinds of seeds used in activity D-3. Your 4 graph lines will have different slopes (and not superimpose) if your seeds have different sizes.

Three data points are required for each table even though 2 points define a straight line. The extra point verifies the validity of the other 2 if it falls on the same straight line.

Students are free to find the mass of any number of seeds they like, keeping in mind that their total mass must not exceed 20 grams, the upper limit of the graph. The larger the seed sampling they choose, the more accurate will be their resulting graph line.

Check Point

1. Data will vary depending on the kinds and quantities of seeds weighed. Here is one result:

a. Beans				b. Corn			
NUMBER OF SEEDS	0	20	40	NUMBER OF SEEDS	0	30	60
MASS IN GRAMS	0	5.24	10.50	MASS IN GRAMS	0	3.82	7.62

c. Lentils				d. Rice			
NUMBER OF SEEDS	0	40	80	NUMBER OF SEEDS	0	50	100
MASS IN GRAMS	0	1.98	3.98	MASS IN GRAMS	0	.78	1.58

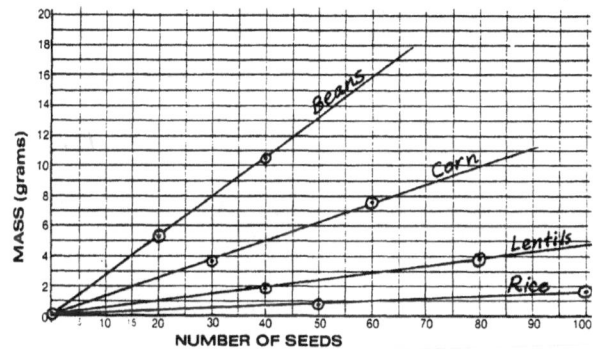

2. Heavier seeds have graph lines that are steeper than the graph lines of lighter seeds.

3a. Seeds that float on air accumulate very little mass as you increase their numbers. This produces an extremely flat, nearly horizontal graph line.

3b. Seeds that are heavier than coconuts, by contrast, increase their mass enormously with the addition of each new seed. This produces an extremely steep, nearly vertical graph line.

E. DIFFERENT DIMENSIONS

E-1 length
E-2 area
E-3 volume
E-4 which dimension?
E-5 bottle cap measure

E-6 dry measure / liquid measure
E-7 all kinds of measure
E-8 measuring with water
E-9 seeds in a can
E-10 salt and sugar

Before students can learn *how* to measure, they need to understand *what* they are measuring. To develop a true sense of space, to conceptually distinguish between length, area and volume, students need to experience the difference between one, two and three dimensions in concrete terms.

These experiments provide that concrete experience. Students make real centimeter cubes from paper and tape. They line them up to measure distances; cover surfaces to measure area; fill boxes to measure volume.

By definition, the whole metric system conceptually derives from a centimeter cube. Your students will build this cube then determine that it holds exactly 1 gram of water. Centimeters (single, squared or cubed), milliliters and grams all tie together in an interrelated web of concrete experience.

━━ EVALUATION ━━

Each question evaluates a single activity from DIFFERENT DIMENSIONS as numbered. Use any combination to frame a formal exam or an informal review. Copy these questions on your blackboard, construct your own ditto master, or photocopy the questions while masking out the rest of the page. Evaluate in ways that suit your own teaching style, enabling your students to learn and enjoy science.

Questions

E-1
Use your metric cube models to measure the width of your table to the nearest whole centimeter.

E-2
Use your metric cube models to measure the area of your desk to the nearest whole sq cm.

E-3
Use your metric cube models to estimate the volume of space under your desk.

E-4
This box is marked off in units of length called "units." Find each measure:

FRONT

a. height of box
b. area of front
c. volume of box

E-5
A small glass jar hold 50 grams of oil when filled to the brim. How would you use it to measure out 1000 grams of oil?

E-6
How many 100 ml cups fill this box? Show your math.

100 ml
20 cm
5 cm   5 cm

E-7
A level teaspoon holds 5 ml of water. What is the mass of 100 teaspoons of water?

E-8
Explain how you would use a 125 ml cup and a 50 ml cup to measure out 75 ml of water.

125 ml   50 ml → 75 ml

E-9
You can pour about 45 grains of a certain kind of rice into a metric cube. Estimate how many of these grains of rice would fill a liter jar (1000 ml).

E-10
The graph line for water is the one labeled b. Which graph line represents wood? Explain how you know.

a.   b.   c.
MASS
VOLUME

Answers

E-1
(Students should mark off 10 cm spaces along the width of their desks until they find the total number of cube lengths that fit. They must express their measurement as *both* a number and a unit.)

E-2
(Students should mark off 10 cm spaces along the width and length of their desks until the find the length of each dimension. Then they can multiply these numbers together to find the approximate surface area in sq cm.)

E-3
(Students should measure the width, length and height of their desks then multiply these numbers together to find the volume in cu cm.)

E-4
a. height = 10 units
b. area = 80 square units
c. volume = 560 cubic units

E-5
Measure out 1000 grams of oil by dipping the small glass jar into the oil 20 times.
1000 grams x 1 bottle/50 grams
= 20 bottles

E-6
First, find the volume of the box.
V = 5 cm x 5 cm x 20 cm =
500 cu cm = 500 ml
500 ml x 1 cup/100 ml = 5 cups.

E-7
100 tsp x 5 ml/tsp x 1 g/ml = 500 g

E-8
Fill the 125 ml cup full of water, then empty it into the empty 50 ml cup. This leaves 75 ml of water remaining in the larger cup.

E-9
1000 ml x 45 rice grains/ml =
45,000 rice grains

E-10
Line c. It represents a substance (like wood) that is lighter than an equal volume of water.

SEQUENCING

DIFFERENT DIMENSIONS can only be scheduled after students complete their gram balances in D. It is desirable, though not mandatory, to follow up this comparative study of metric length, volume and mass with a concentrated review of the metric relationships in F.

Related Activities: D—E---F---G

MATERIALS

Here is everything your students will use for the next 10 activities on DIFFERENT DIMENSIONS. Materials printed in normal type are part of the core 15-things-in-a-box inventory that support all 100 activities. Materials printed in *italics* are additional local materials that you provide or ask your students to bring from home. Pencil and paper are already assumed and therefore unlisted. Each item is numbered with the activity where it is first used.

(E-1) Scissors.
(E-1) Clear tape.
(E-5) Paper clips.
(E-5) *Bottle caps.*
(E-5) Masking tape.
(E-5) Aluminum foil.
(E-5) Size-D dry cells. (Dead ones are OK.)
(E-5) A source of *water* plus several large containers for *easy* distribution.
(E-5) Equal arm balances plus gram masses from the "D" series of activities.
(E-6) *Small glass jars.*
(E-7) *Lined notebook paper.*
(E-8) *Medium-sized cans.*
(E-8) *Large glass jars.* Each needs to hold at least 1 liter.
(E-9) *Small seeds.* Use lentils, if possible. Or substitute large-grained rice. See "Preparation" in teaching notes E-9.
(E-10) *Refined sugar and salt.*

FURTHER STUDY

Use problems like these plus "extension" ideas in DIFFERENT DIMENSIONS to lead your students beyond worksheet activity into original research and investigation. Each discovery leads to more questions, deeper questions, better questions than these. Answering them is what good science is all about.

Read "*Flatland*" by Edwin A. Abbott. Explore the fascinating 2-dimensional world of the flatlanders from your own 3-dimensional perspective. Use this experience to stretch your mind into 4 dimensions!

How big is our planet? How far around? What is the capacity of our oceans? Write a book of Earth facts. Include units of measure that are easy to visualize and understand.

How did people measure distance before rulers were invented? Study the origin of terms like cubit, span, fathom and inch.

LENGTH

1 | Carefully cut out the *smaller* 7-square pattern.

Cut all around the outside.

2 | Fold along the lines, then tape into a cube.

FOLD   TAPE

3 | *Every* edge on this metric cube measures 1 centimeter.

I'm small, but the whole metric system counts on me!

Use your cube to measure the length of each line:

a. |—|

b. |————|

c. |————————|

4 | Carefully cut out the larger pattern, then fold and tape it into a 10-cube model. Use it to measure the length of each line:

USE US LIKE A RULER.

| 1 | 2 | 3 | 4 | 5 | 6 | 7 | 8 | 9 | 10 |

WRITE YOUR NAME ON EACH MODEL.

d. |————————————|

e. |——————————|

f. |——————|

g. |————————————————|

h. |——————————————|

i. |————————————|

j. |————————————|

k. |————————————————|

l. |————————|

CUT OUTS E-1

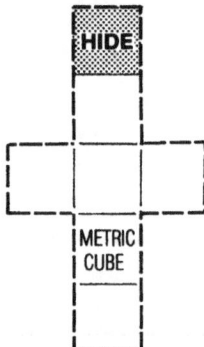

HIDE

METRIC CUBE

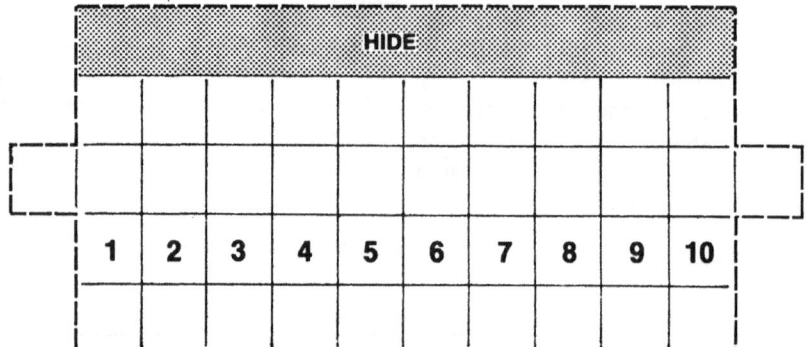

HIDE

| 1 | 2 | 3 | 4 | 5 | 6 | 7 | 8 | 9 | 10 |

Objective

To measure distance with a metric ruler. To recognize the metric cube as a basic unit of metric expression.

Introduction

Cut out the centimeter cube pattern from a *Student Cutouts Booklet*. Fold it together and tape as instructed. Place it on a pedestal — an inverted can will do — in full view of your entire class. Now introduce this modest little cube as you would a distinguished visitor. Be dramatic.

Ladies and Gentlemen, may I present METRIC CUBE — the fundamental building block of the entire metric system — your key to understanding meters, liters and grams!

● *100 of these cubes placed end to end make 1 METER:* Draw a line on the blackboard to represent approximately one meter. Illustrate how 100 metric cubes fit end to end along this line.)

● *1000 of these cubes filled with water occupy 1 LITER:* Turn on an imaginary faucet. Rapidly fill and empty the metric cube into a liter or quart jar (real or imaginary), counting as you go. The jar is full when you reach 1000.

● *1 of these cubes filled with water weighs 1 GRAM:* Fill the cube one more time from your imaginary faucet. Balance it in one hand against a 1 gram mass (from activity D-7) in your other hand. Illustrate with appropriate hand motions how they weigh the same.

The above drama summarizes fundamental concepts about the metric system in an entertaining and visual way. Your students don't have to understand every detail in your presentation. Simply being exposed to these ideas for the first time will make them easier to understand later on as students work through each activity.

Lesson Notes

1-4. Both metric models will be used in many activities to come. Students should build both with care, identify them by name or personal initials, and set them aside in a safe place for future use.

Students will complete this activity rapidly. Be prepared to move them right into the next as soon as they finish this one.

3-4. Students must express each measurement as both a number *and* a unit. When you see numbers written without units, refer to them as so many "oranges" or "bananas" or something ridiculous. A number by itself is unspecified. It can refer to anything. So call it anything.

Check Point

These measurements are *not* expressed as accurately as they could be. Each length is more properly estimated to the nearest tenth centimeter, not rounded off. Your class will learn more about this later when they study significant figures in the "G" series of activities.

a. 1 cm
b. 3 cm
c. 5 cm
d. 8 cm
e. 6 cm
f. 4 cm
g. 10 cm
h. 13 cm
i. 7 cm
j. 9 cm
k. 14 cm
l. 11 cm

AREA

Every surface on this metric cube measures one *square* centimeter.

length x width = AREA

1 cm x 1 cm = 1 sq cm

Use your metric models to find the *area* of each rectangle below. Then check each answer by multiplying.

FIRST count squares, *then use math to check.*

square cm
1 square cm | 1 square cm

BOX	Count how many squares fit.	Multiply to check your answer. (length x width = area)
a.		
b.		
c.		
d.		
e.		
f.		

WRITE ALL UNITS

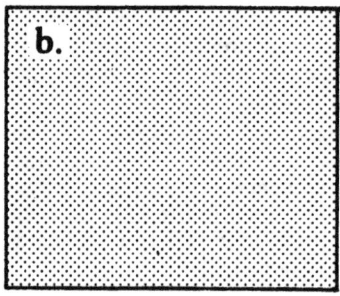

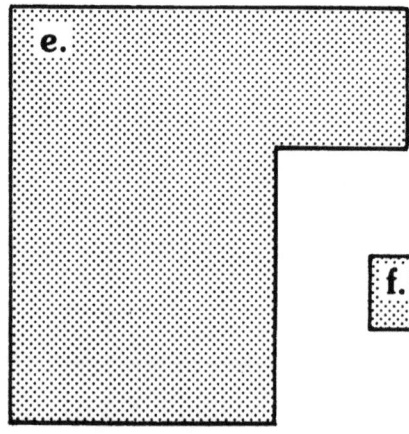

a.

b.

c.

d.

e.

f.

Objective

To develop a concrete understanding of area. To learn how to calculate area and express it as squared measure.

Introduction

Draw a 5 x 4 grid of squares on your blackboard. Make the squares large enough to see from a distance. Since the length of each square is too long to be called a a centimeter, call it a "unit" instead. Show how to find the area of the entire rectangle by (a) counting all the square "units" (b) by multiplying length by width.

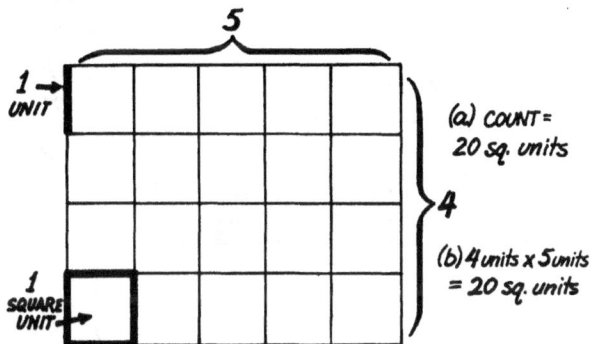

Finding area is a matter of counting squares. To figure the total number of squares, count how many are in a single row, then multiply by the total number of rows. Figuring area is that simple.

Lesson Notes

Students need to work across the table, not down. They should count the squares in one box and confirm by multiplying that they have found the correct area, before proceeding to the next box on the next line. This process reinforces the important idea that area is 2-dimensional space, a product of 1-dimensional width and length.

Remind students once again to write units with each measure. Because the numbers in this table are arranged in columns, one above the other, some may substitute ditto marks instead. This is not a good practice. It is too easy to write ditto marks without thinking — and thus without learning. The extra effort required to write units (they're only abbreviations) is good medicine worth taking.

Check Point

(Calculate (d) and (e) by dividing each large area into 2 smaller rectangles.)
a. 24 sq cm — 8 cm x 3 cm = 24 sq cm
b. 20 sq cm — 5 cm x 4 cm = 20 sq cm
c. 22 sq cm — 2 cm x 11 cm = 22 sq cm
d. 46 sq cm — 6 cm x 7 cm + 1 cm x 4 cm = 46 sq cm
e. 28 sq cm — 6 cm x 2 cm + 4 cm x 4 cm = 28 sq cm
f. 1 sq cm — 1 cm x 1 cm = 1 sq cm

NAME: CLASS:

Different Dimensions **E-3**

VOLUME

This metric cube occupies
one *cubic* centimeter of space.

Use your metric models to find the *volume*
of each box below. Check each answer
by multiplying.

My volume is
1 cu cm.

First count cubes.

Then use math
to check your
answer.

1 cm x 1 cm x 1 cm = 1 cu cm
(length)   (width)   (height)   (volume)

BOX	Count how many cubes fit.	Multiply to check your answer. (length x width x height = volume)
a.		
b.		
c.		
d.		
e.		
f.		

WRITE ALL UNITS.

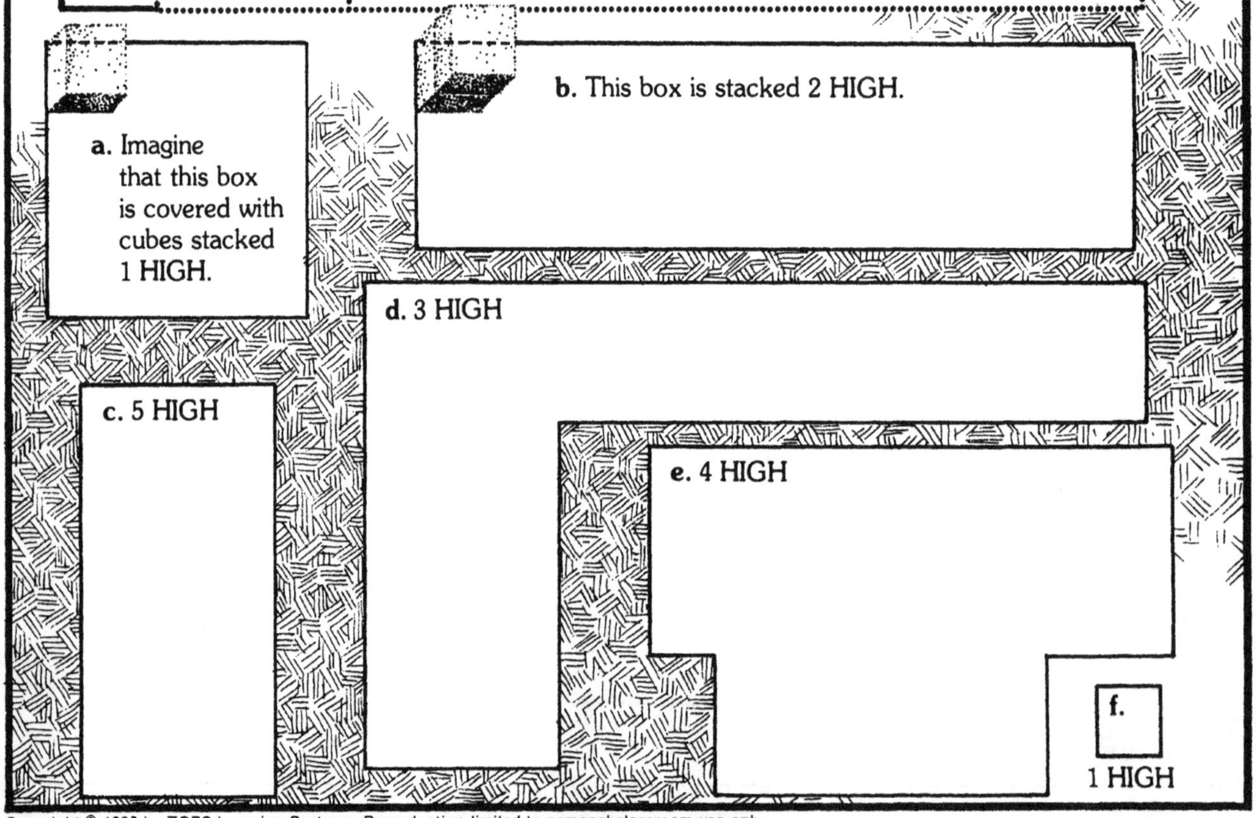

a. Imagine that this box is covered with cubes stacked 1 HIGH.

b. This box is stacked 2 HIGH.

d. 3 HIGH

c. 5 HIGH

e. 4 HIGH

f.
1 HIGH

Objective

To develop a concrete understanding of volume. To learn how to calculate volume and express it as cubed measure.

Introduction

Draw a 5 x 4 grid of squares on your blackboard. Make the squares large enough to see from a distance. Now extend this grid backward into 3 dimensions, turning the squares into cubes. Call the length of each cube 1 "unit". Show how to find the volume of the entire block by (a) counting all the "cubic units" (b) by multiplying length by width by height.

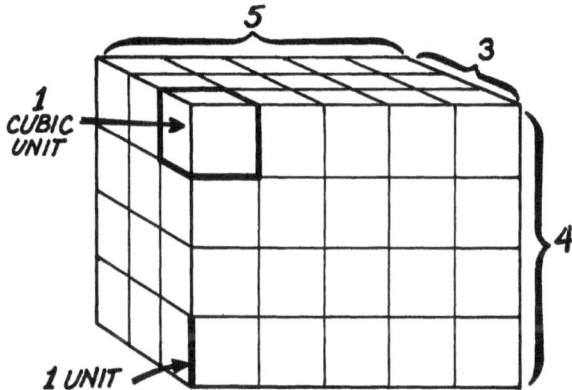

(a) Count = 60 cubic units
(b) 5 units x 4 units x 3 units = 60 cu units

Finding volume is a matter of counting cubes. To figure the total number of cubes, multiply two dimensions together to find the number of cubes in a single layer, then multiply that result by the total number of layers. Figuring volume is that simple.

Lesson Notes

As before, students need to work across the table, not down. Using their metric models and a little imagination, they "fill" each figure and count the cubes, then they confirm by multiplying that they have found the correct volume, before proceeding to the next box on the next line. This process reinforces the important idea that volume is 3-dimensional space, a product of 1-dimensional width, length and height.

When confronted with a row of three numbers, some students may confuse operations, taking their sum instead of their product. To find the product first multiply two dimensions together, then multiply the resulting product by the third dimension.

All admonitions about writing units with each measurement apply here as well (and in worksheets to come). Unitless measurements don't mean a thing. Challenge someone who forgets to write units to "give you five". No, it's not a handshake unless you say it is. And then you've specified the unit!

Extension

Estimate how many metric cubes would fill your classroom. Is it more than a million? More than a billion? Show your math.

A 10 by 6 meter room with a ceiling 3 meters high holds nearly a fifth of a billion metric cubes!

$$(10 \times 10^2 \text{ cm})(6 \times 10^2 \text{ cm})(3 \times 10^2 \text{ cm}) =$$
$$180 \times 10^6 \text{ cu cm} =$$
$$.180 \times 10^9 \text{ cu cm} = .18 \text{ billion cu cm}$$

Check Point

a. 16 cu cm — 4 cm x 4 cm x 1 cm = 16 cu cm
b. 66 cu cm — 11 cm x 3 cm x 2 cm = 66 cu cm
c. 90 cu cm — 3 cm x 6 cm x 5 cm = 90 cu cm
d. 117 cu cm — 12 cm x 2 cm x 3 cm + 3 cm x 5 cm x 3 cm = 117 cu cm
e. 136 cu cm — 8 cm x 3 cm x 4 cm + 5 cm x 2 cm x 4 cm = 136 cu cm
f. 1 cu cm — 1 cm x 1 cm x 1 cm = 1 cu cm

WHICH DIMENSION?

1 Cut, fold, and tape the box pattern.

Then use your metric models to complete this table.

EXAMPLE

Measure what?	Length, Area, or Volume?	Answer? Units?
h	*length*	*2 cm*
w		
top		
front		
whole box		
l		
side		
d		

2 Look around your room. Find 3 examples of each.

LENGTH (1 dimension)	AREA (2 dimensions)	VOLUME (3 dimensions)
a. *the edge of a door.*	d.	g.
b.	e.	h.
c.	f.	i.

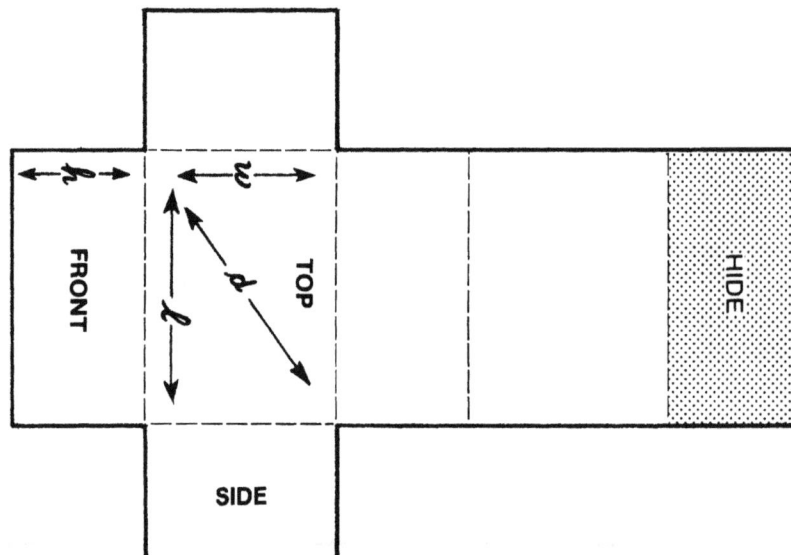

CUTOUTS E-4

FRONT   *h*   *w*   TOP   *d*   *l*   HIDE

SIDE

Objective

To measure a box in one, two, and three dimensions. To distinguish between length, area and volume.

Introduction

A great way to introduce this lesson and review the 3 dimensions studied so far is to physically act them out! You go through the motions while your class identifies each dimension in unison reply.

Once your students are familiar with each visualization, reverse the procedure. You call out length, area, or volume, and allow your class to respond kinesthetically. The mind absorbs more rapidly and more completely when the whole body is involved in the learning process. Learning becomes just plain fun.

YOU CALL OUT . . .	YOUR CLASS ACTS OUT . . .
length	length
area	area
volume	volume
side x side	area
area x side	volume
side x side x side	volume
centimeters	length
square centimeters	area
cubic centimeters	volume
1-dimension	length
2-dimensions	area
3-dimensions	volume

Lesson Notes

2. All objects have 1, 2 and 3 dimensions. Lengths, areas and volumes therefore, are to be found everywhere in the classroom. Your students will likely confine their observations, however, to shapes they have studied — straight lines, rectangles and blocks.

Reserve at least one of these boxes to use in your introduction to activity E-6. The rest may be discarded.

Check Point

1.
h	length	2 cm
w	length	3 cm
top	area	12 sq cm
front	area	8 sq cm
whole box	volume	24 cu cm
l	length	4 cm
side	area	6 sq cm
d	length	5 cm

2a. the edge of a door
 b. a crack in the floor
 c. a line on a piece of paper
 d. a window surface
 e. the face of a book
 f. a piece of paper
 g. a box
 h. a text book
 i. the entire room

BOTTLE CAP MEASURE

1 Bend out a paper clip to a right angle.

Stick it to the bottom of a bottle cap with masking tape.

Write your name on your cap.

2 Cut out a square of aluminum foil 10 cm on each side.

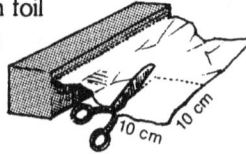

10 cm   10 cm

Then fold it up around a dry cell.

3 Push down the jagged edges.

Take out the dry cell, and slide a paper clip onto one side. This makes a "bucket."

4 Pour a bottle cap full of water into your bucket. Check for leaks, then pour the water back out.

DRY!

5 Get your balance. Replace the *left* basket with this foil bucket.

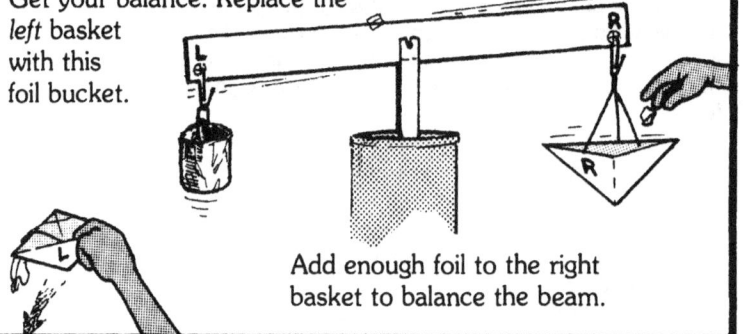

Add enough foil to the right basket to balance the beam.

6 Dip a full bottle cap of water into your balance bucket.

Find the mass of this water.

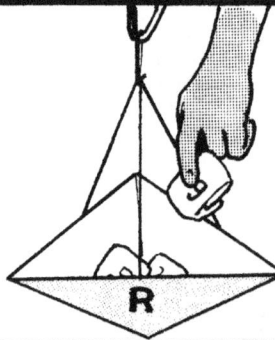

Do this 3 times to find the average mass of water.

| TRIAL 1: |
| TRIAL 2: |
| TRIAL 3: |
| AVERAGE: |

7 If you wanted to fill a glass with exactly 100 grams of water, how would you do it?

8 Hang the basket on your beam again. Save your bucket and bottle cap.

Objective

To develop a metric volume standard for measuring out small amounts of water.

Introduction

None required.

Lesson Notes

2. Students should measure the foil with their 10 cube model constructed in activity E-1.

4-6. The foil bucket weighs considerably more than the paper pan it replaces. Because the rider is not heavy enough to compensate for this weight shift, it is not used in the centering process. Instead, additional pieces of foil are added to the right paper pan until the beam recenters. These foil pieces must *remain* in the right pan until the final mass of the water is determined in step 6.

Besides checking for leaks, there is another good reason for adding water to this "bucket" (step 4). Centering the balance to wet bucket conditions (step 5) makes it ready to measure water (step 6) without further adjustments.

6. To take uniform water samples, dip the bottle cap into any large container of water that is free of all traces of soap. Then lift it straight out.

Surface tension may make the water slightly rounded on top. Pour all of this water (without spilling) into the foil bucket. Touch the last drop to the inside of this bucket as well, but don't shake the bottle cap.

If any particular trial seems unusually high or low, take another sampling. Repeat as often as necessary until you get consistent results. Samples should not differ from one another by more than a tenth of a gram or so.

The masking tape that holds the bottle cap to its paper clip handle is not waterproof. If it comes loose, dry everything thoroughly and retape. Rewrite your name on the bottle cap if it came off with the tape.

8. Both the bucket and bottle cap should be saved with the other metric models. Since the capacities of bottle caps will vary, they must each be identified with names so that students can consistently use the *same* ones in later activities.

Check Point

6. Answers will vary depending on the volume of the bottle cap. Here is one result:

$$\begin{aligned} \text{trial 1: } & 3.40 \text{ grams} \\ \text{trial 1: } & 3.48 \text{ grams} \\ \text{trial 3: } & 3.44 \text{ grams} \\ \text{average: } & 3.43 \text{ grams} \end{aligned}$$

7. Add 29 bottle caps of water to the glass. Since each cap contains 3.43 grams, this would yield about 100 grams of water.

100 grams x 1 cap/3.43 grams = 29 caps of water

DRY MEASURE / LIQUID MEASURE

1 CUTOUTS E-6

Cover the figure with clear tape, then cut around the outside.

Tape, then cut out.

2 Fold up all 4 flaps to make a box. Seal the edges to hold water.

FOLD OVER

Tape top to bottom.

TRIM

3 There are 2 ways to measure volume. . . . . . Understand this, and you can fill in the table.

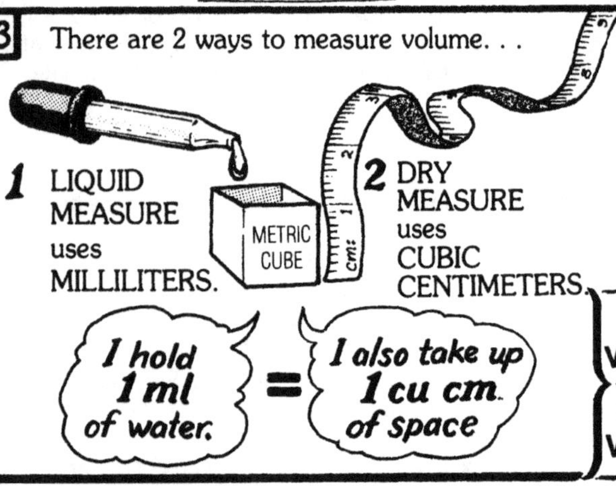

1 LIQUID MEASURE uses MILLILITERS.

METRIC CUBE

2 DRY MEASURE uses CUBIC CENTIMETERS.

*I hold **1 ml** of water.* = *I also take up **1 cu cm** of space*

ALWAYS write units with each answer!

YOUR METRIC MODELS

	a.	b.	c.
DRY VOLUME?			
LIQUID VOLUME?			

4 Bend out a paper clip at a right angle.

Tape it to the side of the box you made.

5 Measure 100 ml of water into a small jar. Mark the water level on masking tape.

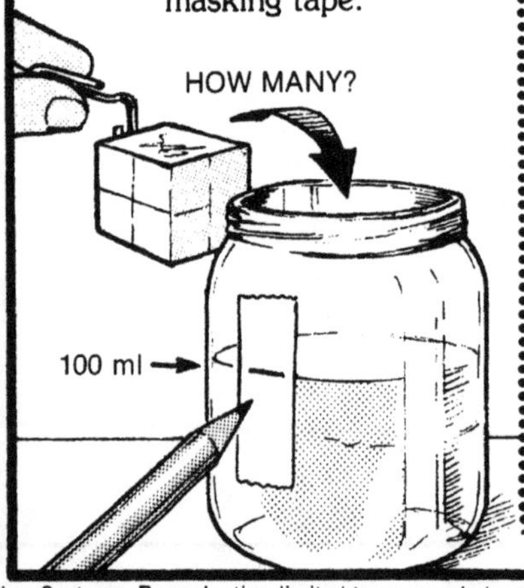

HOW MANY?

100 ml →

Tell how you measured the water.

Tape your name to the box and jar, and save them.

Objective

To find volumes using liquid measure. To calibrate a glass jar at the 100 ml level.

Introduction

Your students have already learned to calculate volume with a ruler by multiplying length by width by height. This method is known as "dry measure", because no water is involved. Dry measure works fine on boxes. It also works for cylinders, cones, pyramids, spheres and other regular solids if you happen to remember the volume formulas involved. But the mathematics of dry measure gets too complicated for irregular containers. This is a job for liquids, substances that take the shape of any container you put them in, no matter how irregular.

To introduce the idea of liquid measure, hold an imaginary demonstration. Assemble the following props to use in this order: 1-cube model and 10-cube model (from activity E-1); an empty beverage can or bottle with its liquid volume stated on the outside of the container (likely 355 ml); the 2 x 3 x 4 cm box saved from activity E-4.

First hold up the 1-cube model. Fill it with imaginary water from an imaginary faucet. (Never mind that it has a lid. Pretend it isn't there.) Your students should observe that 1 milliliter of water isn't much. It only fills, after all, a small centimeter cube. Write the relationship between milliliters and centimeters on your blackboard (1 ml = 1 cu cm).

Next display the 10-cube model plus an empty 355 ml beverage container. Fill the 10-cube model with imaginary water and ask how much water it contains (10 ml). Announce that you will now determine how many 10 ml portions fill the beverage can. Pretend to add water, counting as you go: 10 ml, 20 ml, 30 ml . . . Count by tens until your reach 350 ml, then announce that there is room for only a half portion more (5 ml). The container overflows at 355 ml. Ask students who don't believe you to read the liquid volume on the side of the container! Finding liquid volume is that simple: just add known volumes of water (milliliters) counting as you go. When you get to the top, stop counting. That's the volume of your container.

Finally, hold up the 2 x 3 x 4 cm box. Write its dimensions on your blackboard. Ask your class its dry volume (24 cu cm) and liquid volume (24 ml). Discuss how many boxes filled with imaginary water will fill the beverage container:

355 ml x 1 box/24 ml = 15 boxes (approx).

Lesson Notes

1. This pattern folds into a box that actually holds water. Covering it with clear tape waterproofs the paper long enough to complete the experiment.

2. The box is most easily formed by sharply creasing the bottom edge of all 4 flaps. This places the flaps in a near-upright position where they can be clear-taped together at vertical right angles.

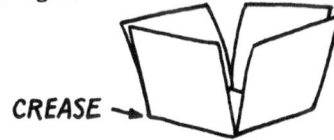

CREASE →

The edges and corners don't require a perfect seal. Used as a dipper in step 5 to transfer water into a glass jar, this box only has to hold water a few seconds at a time.

Don't worry if some of the measuring boxes are less than perfectly formed. The experimental process is more important than any particular calibration. Do your students understand how water is used to measure volume? This is most important.

Extension

If you have a graduated cylinder or buret, try this demonstration. Assemble an assortment of containers both large and small. Ask your class to estimate the volume (in milliliters) of one particular container. Write guesses on the blackboard. Then find the volume experimentally (using the graduate or buret) to see who estimated closest. Repeat using other containers until your class develops a "feel" for milliliters.

If your class enjoys competition, divide into teams. Subtract estimated volumes from measured volumes to see which team scores the lowest total cumulative error.

Check Point

3.

	a.	b.	c.
DRY VOLUME?	1 cu cm	10 cu cm	8 cu cm
LIQUID VOLUME?	1 ml	10 ml	8 ml

5. The 8 cu cm box holds 8 ml. To measure out 100 ml, fill it 12.5 times with water and empty it each time into the small jar. The water level in this jar will reach to 100 ml.

100 ml x 1 box/8 ml = 12.5 boxes

ALL KINDS OF MEASURE

1 If you fill a metric cube with water, it will balance one gram. . .

Understand this, and you can fill in the table.

My WATER MASS equals ONE GRAM.

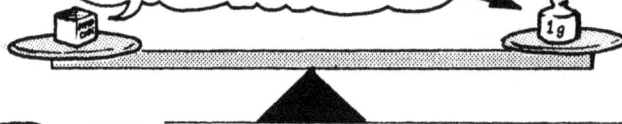

Work DOWN each column, not across.

	a.	**b.**	**c.**	**d.**	**e.** 100 ml
MASS?					
LIQUID VOLUME?					
DRY VOLUME?					
LENGTH?					
AREA? (all outside surfaces)					

WRITE ALL UNITS

2 How many bottle caps of water will reach the 100 ml tape level?

a. Predict first. Explain your reasoning.

b. How close was your prediction?

3 Make a 1 liter box: tape together 4 paper squares that measure 10 cm on each side.

10 cm

1 liter

LABEL your box.

If your liter box could hold water . . .

a. How many 100 ml volumes would fill it?

b. What is the mass of this much water?

Objective

To distinguish between 5 different units of measure.

Introduction

None required.

Lesson Notes

1. This table reviews 5 different measuring units in terms of the metric models constructed thus far. Assume that each of the models is made only of water; neglect the mass of the container.

Notice that students are directed to work down the table, not across. Working down each column forces students to continuously interchange units, and thus better understand how they are related.

Length and area are omitted in column (d) and (e), not because these models don't possess these physical quantities. Rather they are harder to specify or more difficult to calculate.

3. Paper squares are easy to produce on notebook paper: Let a corner define 2 sides of the square. Then make only 2 additional cuts — one parallel with lines in the paper, and the other perpendicular to these lines.

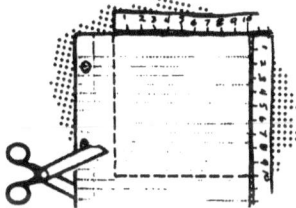

These 4 squares should be joined by small pieces of clear tape, perhaps 2 to a corner as illustrated. Students tend to waste tape in this step, applying long strips along the entire length of each corner.

Check Point

1.

	a.	b.	c.	d.	e.
MASS?	1 g	10 g	8 g	3.43 g	100 g
LIQUID VOLUME?	1 ml	10 ml	8 ml	3.43 ml	100 ml
DRY VOLUME?	1 cu. cm	10 cu cm	8 cu cm	3.43 cu cm	100 cu cm
LENGTH?	1 cm	10 cm	2 cm		
AREA? (all outside surfaces)	6 sq. cm	42 sq cm	20 sq cm		

2a. Here is a prediction based on a bottle cap capacity of 3.43 ml.   100 ml x 1 cap/3.43 ml = 29 caps of water

b. 29 bottle caps of water came reasonably close to the 100 ml line.

3a. A liter cube measures 10 cm x 10 cm x 10 cm. This multiplies to 1000 cu cm or 1000 ml. Ten 100 ml jars of water are required to fill it up:
    1000 ml x 1 jar/100 ml = 10 jars

b. 1000 ml of water weighs 1000 grams or 1 kilogram.

NAME: CLASS:

MEASURING WITH WATER

1 Use your metric models to find the volume of a can.

How much will your can hold?

Explain how you calculated your result.

2 Stick a strip of masking tape to a large glass jar like this.

3 Add water to this jar, 100 ml at a time. Mark the water level after each addition, up to 1000 ml.

I'm CALIBRATING the jar.

4 Use your calibrated liter jar to measure the volume of your can again.

Explain how you found its volume this time.

5 Compare your results in steps 1 and 4.

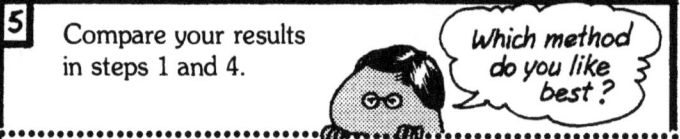

Which method do you like best?

Tape your name to the can, and save it.

Objective

To find the volume of a can using different instruments of liquid measure. To calibrate a jar in 100 ml increments up to 1 liter.

Introduction

None required.

Lesson Notes

1. Students should add measured amounts of water up to but not exceeding the top of the can.

LEVEL: TOO FULL:

5. The jars can be returned to storage at this time. The cans will be used in the next activity.

Check Point

1. Here are 2 possible ways to measure the volume of a can:

a) Fill the can nearly full with 100 ml portions, then use the smaller 8-ml box to fill the can the rest of the way. Find the sum of all the volumes you add.

Example:

$$4 \times 100 \text{ ml} = 400 \text{ ml}$$
$$3 \times 8 \text{ ml} = 24 \text{ ml}$$
$$.5 \times 8 \text{ ml} = \underline{4 \text{ ml}}$$
$$428 \text{ ml}$$

b) Fill the can full, with 100 ml portions. Measure the volume of water that is left over from your last portion with your 8-ml box and subtract this amount from the total volume.

Example:

$$5 \times 100 \text{ ml} = 500 \text{ ml}$$
$$9 \times 8 \text{ ml} = \underline{-72 \text{ ml}}$$
$$428 \text{ ml}$$

4. Fill the calibrated jar up to the the 1 liter (1000 ml) mark. Then pour this water into the can until it is even with the top. Subtract the volume that is left in the jar from 1000 ml to find the volume of water you added to the can.

Example:

$$\text{liter jar} = 1000 \text{ ml}$$
$$\text{water left in jar} = \underline{-580 \text{ ml}}$$
$$420 \text{ ml}$$

5. The values in steps 1 and 4 should be roughly equal. Expect small differences due to normal measuring error. Your students will likely prefer using the calibrated liter jar because it's easier and faster to use.

SEEDS IN A CAN

1 Scoop up a level capful of seeds and count them all.

LEVEL!

...47, 48, 49 ...

MAKE PILES OF 10.

SCOOP

COUNT

Do this 3 different times to find an average.

TRIAL 1:

TRIAL 2:

TRIAL 3:

AVERAGE:

2 You know. . .

- The volume of the small bottle cap and the large can.

- How many seeds fill the bottle cap.

THIS I KNOW...

SHOW YOUR MATH.

Use this information to estimate the number of seeds that will fill your tin can.

How many?

3 Estimate the number of seeds in your can *again*. This time use your 8 ml box.

ESTIMATE WITH THIS!

Show your work here.

4 Which estimate do you think is most accurate? Why?

Objective

To estimate the number of seeds in a can by comparing volumes.

Preparation

Choose appropriately-sized seeds for this experiment. They should be small enough to fill confined volumes (the metric bottle cap and 8 ml box), yet not so small that they present unmanageable numbers for counting. Lentils are an ideal size. Rice grains are OK *if* you can find large ones. Pour them into an 8 ml box and see how long it takes to count them. If the count seems too tedious, select a larger seed.

Lesson Notes

2-3. Long division and multiplication are required here. If your students don't need the math practice (most do), calculators will speed up these calculations.

4. If your students are curious to know the exact number of seeds in the can, ask them to go ahead and count them! As a cooperative activity, large classes can accomplish this task very quickly. Carefully divide all the seeds in a can among all of your students. Ask them to return the seeds to this can in quantities of 50 while you tally marks on the blackboard. Near the end of the count, students should pool their remaining seeds together to continue delivering groups of 50. Only the last group of seeds will equal less than 50. These should fill the can back up to its rim.

Check Point

(Answers will vary. Here is one result based on lentils.)
1.   trial 1: 54 lentils
       trial 2: 59 lentils
       trial 3: 52 lentils
    average: 55 lentils

2. A 3.43 ml bottle cap holds 54 seeds. Thus the 424 ml can holds: 424 ml x 54 seeds/3.43 ml = 6,675 seeds

3. An 8 ml box holds 158 seeds. Thus the 424 ml can holds:
       424 ml x 148 seeds/8 ml = 7,844 seeds

5. The estimate based on 8 ml is likely more accurate because it is based on a larger sampling of seeds. The most accurate "estimate" would be to count all the seeds in the can. This amounts to a sample size of 424 ml, the volume of the can.

SALT AND SUGAR

1 Fill your metric bottle cap with sugar.

LEVEL!

2 Find the mass of this volume of sugar on your centered balance.

L   R

L   R

Pour sugar here.

Put gram masses here.

3 Fill in this sugar table. Plot your results and label the graph line.

CUTOUTS

E-10

SUGAR		
	VOLUME (ml)	MASS (grams)
ONE bottle cap		
TWO bottle caps		

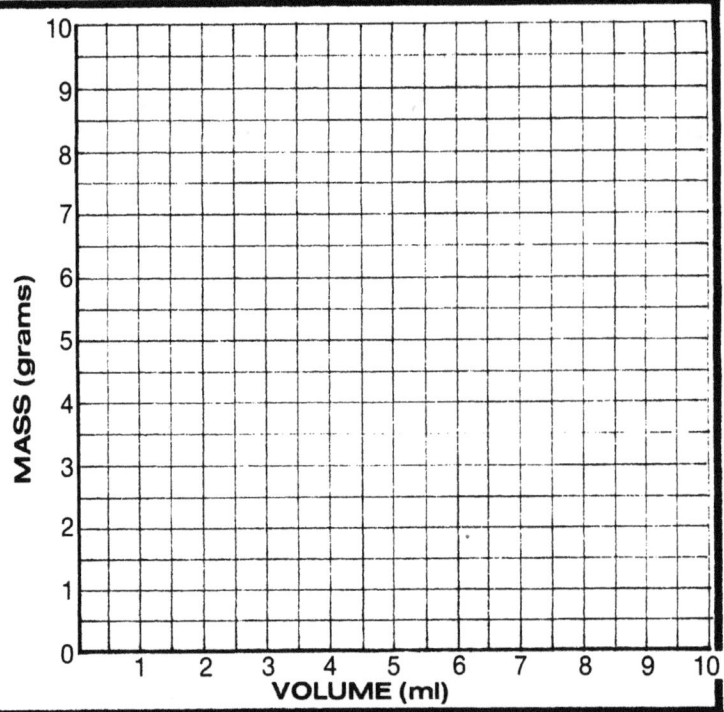

Repeat these steps using salt.

SALT		
	VOLUME (ml)	MASS (grams)
ONE bottle cap		
TWO bottle caps		

MASS (grams)

VOLUME (ml)

4 You know all about metric water cubes. . .

*My **volume** is **1 ml**.*

*My **mass** is **1g**.*

1 cm
1 cm
1 cm

Plot and label a line for *water* on the above graph.

5 Fill your foil bucket with 2 level capsful of *sugar*. Close it tightly.

REMOVE PAPER CLIP

ADD SUGAR   CLOSE TIGHTLY

Your lab partner should do the same using *salt*.

6 Predict whether these will float or sink in water.

WATER?

SUGAR   SALT

a. sugar: **b.** salt:

Test your prediction.

Objective

To correlate the mass of a substance with its floating and sinking characteristics.

Introduction

None required.

Lesson Notes

2-3. Students should recenter their balances after each weighing. Granules of sugar or salt tend to stick between overlapping paper in the weighing pans.

4. The coordinates for water are easy to plot because each ml weighs 1 more gram: (1,1), (2,2), (3,3), . . ., (n,n).

5. Anyone who drops a sugar cube into a cup of coffee knows that it sinks like a stone. But that's only because it rapidly absorbs water. If you first wrap the sugar in foil, it easily floats. It will not sink until water finally penetrates between the layers of foil to soak the sugar.

To keep water from soaking the sugar (or salt) prematurely, the foil should be wrapped tightly around each substance, but not twisted so hard that the foil tears.

6. A graph is *not* a glass of water! Watch out for those who think it is. The faulty reasoning goes something like this: On the graph, salt is above water, while sugar is below. So salt floats while sugar sinks. Wrong!

Extension

The 3 graph lines in this activity show how the mass of each substance increases with volume. The slope of each line — the change in mass divided by the change in volume — is an expression of density.

$$\text{slope} = \frac{\text{change in mass}}{\text{change in volume}} = \text{density}$$

Find the slope of each line to calculate the density of each

substance. Relate your results to the floating and sinking characteristics of salt and water.

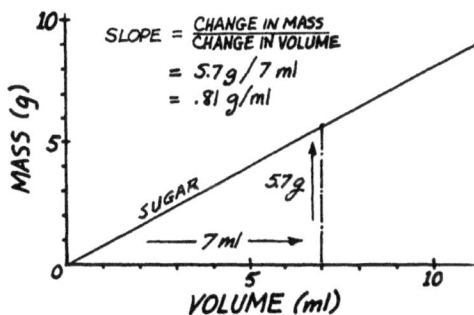

$$\text{SLOPE} = \frac{\text{CHANGE IN MASS}}{\text{CHANGE IN VOLUME}}$$
$$= 5.7\,g / 7\,ml$$
$$= .81\,g/ml$$

slope for sugar = 5.7 g / 7.0 ml = .81 g/ml
slope for water = 7.0 g / 7.0 ml = 1.00 g/ml
slope for salt = 8.8 g / 7.0 ml = 1.26 g/ml

Sugar floats because it is less dense than water. Salt sinks because it is more dense than water.

Check Point

3-4.

SUGAR	VOLUME (ml)	MASS (grams)
ONE bottle cap	3.43 ml	2.82 g
TWO bottle caps	6.86 ml	5.68 g

SALT	VOLUME (ml)	MASS (grams)
ONE bottle cap	3.43 ml	4.32 g
TWO bottle caps	6.86 ml	8.58 g

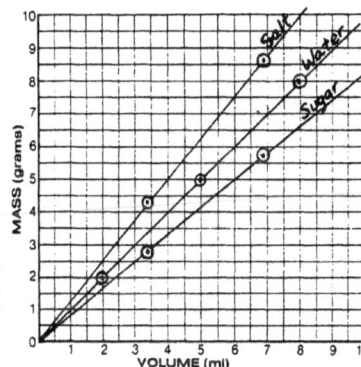

6a. Sugar floats; Salt sinks.
 b. Prediction was correct.

F. TALKING METRIC

F-1 metric language
F-2 layer upon layer
F-3 kilometer ruler

F-4 metric squares
F-5 face-up
F-6 metric rummy

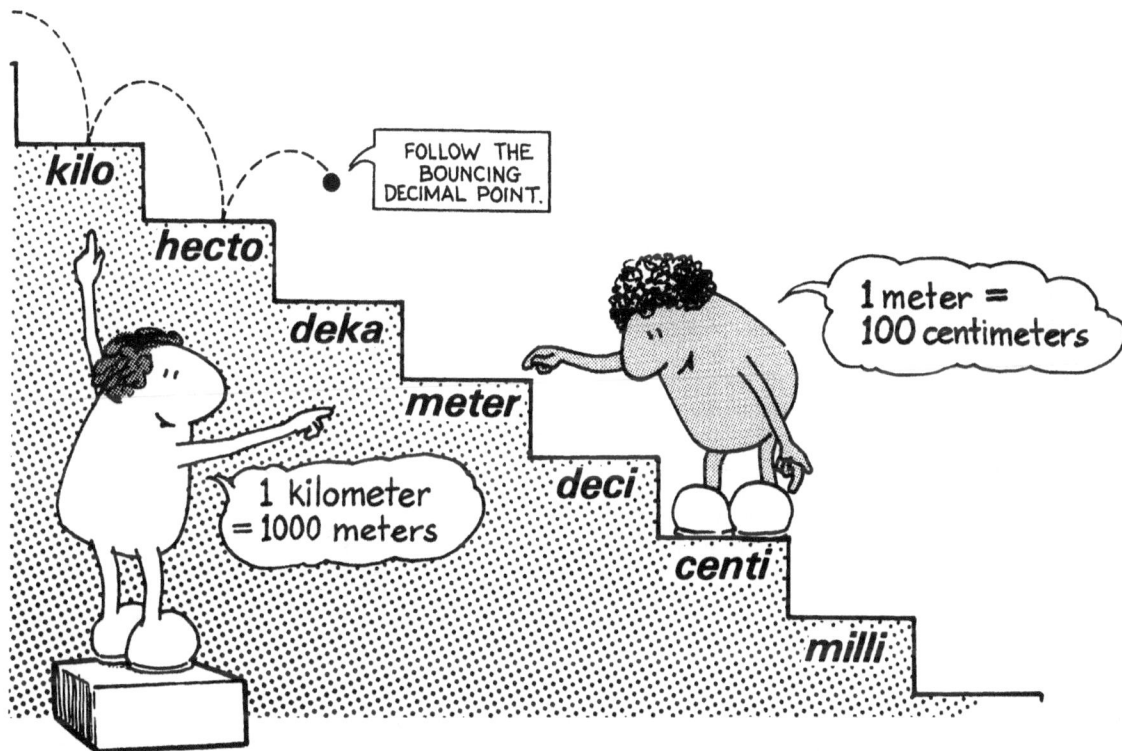

To understand metrics your students must first learn the language. Root words express basic dimensions — meter, liter, gram. These roots are joined to prefixes that denote decimal size — kilo, hecto, deka, deci, centi and milli. Walk a distance of 1000 meters, and you travel a *kilo*meter. Spill a volume of water equal to 1/1000 liter and you've lost a *milli*liter. Eat a mass of cheese equal to 1/100 gram and you've tasted a *centi*gram, probably not enough to swallow.

Learning to interchange these units, then move the decimal point to maintain equality is *not* easy. Be sure

to teach lessons 1 and 3 as a class exercise *before* students attempt the worksheets on their own.

Lessons 5 and 6 help students establish concrete metric associations within a game context. Just as "monkey" brings the image of an animal immediately to mind, "monkey's tail" can establish, once and for all, the idea of 1 meter. These games also require a strong teacher introduction. Demonstrate them to your whole class first, so everyone thoroughly understand the rules, before students individually play each other.

── EVALUATION ──

Each question evaluates a single activity from TALKING METRIC as numbered. Use any combination to frame a formal exam or an informal review: Copy these questions on your blackboard, construct your own ditto master, or photocopy the questions while masking out the rest of the page. Evaluate in ways that suit your own teaching style, enabling your students to learn and enjoy science.

Questions

F-1
a. Complete this pattern, ending with millipapers:

1 kilopaper = 10 hectopapers = 100. . .

b. Complete this pattern, ending with kilopapers:

1 millipaper = .1centipaper = .01. . .

F-2
This line measures 1 "hecto":

1 HECTO ✏

a. draw a "deka" line.

b. draw a "kilo" line.

F-3
Fill in the correct number to complete each equation.

1 meter = _____ cm
1 meter = _____ mm
1 meter = _____ km
1 gram = _____ kg
1 gram = _____ mg
1 liter = _____ ml

F-4
Pair each sentence with the correct metric unit.
a. As much as 4 glasses of water.
b. As far as a 10-minute walk.
c. As wide as a pinhead.
d. As much as 15 drops of water.
e. As heavy as a chicken.
f. As thick as blackboard chalk.
g. As long as your leg.
h. As heavy as 2 paper clips.
i. As heavy as a bread crumb.

gram	millimeter	milligram
milliliter	centimeter	kilometer
kilogram	liter	meter

F-5
Fill in each blank with the most logical metric unit.
a. Maria lives 4 _____ from school.
b. Kim weighs 40 _____.
c. Mohammed is almost 2 _____ tall.
d. A watermelon seed weighs about 60 _____.
e. Jason's hand measures 13 ___ long.
f. Gloria had to swallow 5 _____ of medicine at once.
g. A coin has a mass of about 5 _____.
h. The crack in the sidewalk is 5 _____ wide.
i. Kwasi's car holds 40 _____ of gas.

F-6
Draw a . . .
. . . Circle around each meter;
. . . A square around each cm;
. . . A triangle around each mm.

.01 m   100 cm

.001 km   .001 m   1000 mm

.1 cm   10 mm

Answers

F-1
a. 1 kilopaper = 10 hectopapers =
100 dekapapers = 1,000 papers =
10,000 decipapers = 100,000 centipapers =
1,000,000 millipapers.

b. 1 millipaper = .1 centipaper =
.01 decipaper = .001 paper =
.0001 dekapaper = .00001 hectopaper =
.000001 kilopaper.

F-2
a. (Students should draw a line ten times *shorter* than the sample "hecto" line.)
b. (Students should draw a line ten times *longer* than the sample "hecto" line.)

F-3
1 meter = 100 cm
1 meter = 1000 mm
1 meter = .001 km
1 gram = .001 kg
1 gram = 1000 mg
1 liter = 1000 ml

F-4
a. liter
b. kilometer
c. millimeter
d. milliliter
e. kilogram
f. centimeter
g. meter
h. gram
i. milligram

F-5
a. km
b. kg
c. m
d. mg
e. cm
f. ml
g. g
h. mm
i. l

F-6

.01 m   100 cm

.001 km   .001 m   1000 mm

.1 cm   10 mm

SEQUENCING

TALKING METRIC is a recommended, though not required, sequel to E. Follow up with G if you wish, or return at a later date.

Related Activities: D—E---**F**---G

MATERIALS

Here is everything your students will use for the next 6 activities on TALKING METRIC. Materials printed in normal type are part of the core 15-things-in-a-box inventory that support all 100 activities. Materials printed in *italics* are additional local materials that you provide or ask your students to bring from home. Pencil and paper are already assumed and therefore unlisted. Each item is numbered with the activity where it is first used.

(F-2) Scissors.
(F-2) *Lined notebook paper.*
(F-2) Clothespins.
(F-5) Masking tape. Or substitute blackboard chalk.
(F-5) Paper clips.

FURTHER STUDY

Use problems like these plus "extension" ideas in TALKING METRIC to lead your students beyond worksheet activity into original research and investigation. Each discovery leads to more questions, deeper questions, better questions than these. Answering them is what good science is all about.

There are metric prefixes much larger than "kilo" and much smaller than "milli". Extend this list as far as you can in both directions:

$$kilo = 10^3 \text{ or } 1000$$
$$hecto = 10^2 \text{ or } 100$$
$$deka = 10^1 \text{ or } 10$$
$$10^0 \text{ or } 1$$
$$deci = 10^{-1} \text{ or } .1$$
$$centi = 10^{-2} \text{ or } .01$$
$$milli = 10^{-3} \text{ or } .001$$

Can you list 100 measuring units that are all different? Go for a record! (No fair writing the same unit with a different prefix. If you list "meter" you can't list "centimeter". If you write "inch" you can't write "half inch".)

As a measuring expert, your assignment is to publish a glossary of metric images. A measuring beginner should be able to look up any common metric unit in your glossary and find it expressed in familiar terms. Example: meter — the distance between outstretched hands.

NAME: CLASS:

Talking Metric **F-1**

METRIC LANGUAGE

000001.000000
DIVIDE
MULTIPLY
1000
100
10
1
.1
.01
.001
a KILO paper
a HECTO paper
a DEKA paper
A PAPER
a DECI paper
a CENTI paper
a MILLI paper

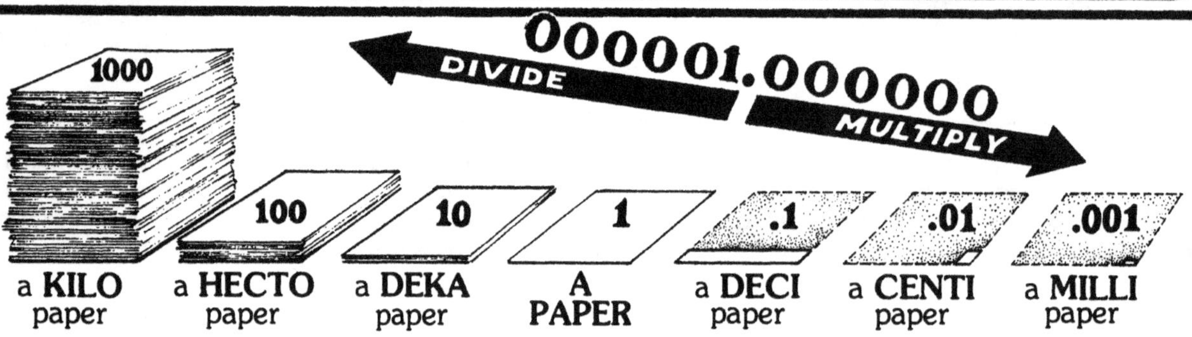

Find the correct number for each box.

a.

1 **KILO**paper =		papers
1 **HECTO**paper =		papers
1 **DECI**paper =		papers
1 **DEKA**paper =		papers
1 **MILLI**paper =		papers
1 **CENTI**paper =		papers

b.

1 PAPER =		**DECI**papers
1 PAPER =		**DEKA**papers
1 PAPER =		**CENTI**papers
1 PAPER =		**HECTO**papers
1 PAPER =		**MILLI**papers
1 PAPER =		**KILO**papers

c.

1 **KILO**paper =		**HECTO**papers
1 **KILO**paper =		papers
1 **KILO**paper =		**CENTI**papers
1 **KILO**paper =		**DEKA**papers
1 **KILO**paper =		**DECI**papers
1 **KILO**paper =		**MILLI**papers

d.

1 **MILLI**paper =		**CENTI**papers
1 **MILLI**paper =		papers
1 **MILLI**paper =		**HECTO**papers
1 **MILLI**paper =		**DECI**papers
1 **MILLI**paper =		**DEKA**papers
1 **MILLI**paper =		**KILO**papers

e.

1 **CENTI**paper =		**MILLI**papers
1 **MILLI**paper =		**CENTI**papers
10 **MILLI**paper =		**CENTI**papers
10 **CENTI**paper =		**MILLI**papers
.1 **CENTI**paper =		**MILLI**papers
.1 **MILLI**paper =		**CENTI**papers

f.

.5 papers =		**MILLI**papers
5 papers =		**MILLI**papers
2 **KILO**papers =		papers
.2 **KILO**papers =		papers
100 **CENTI**papers =		papers
.15 papers =		**CENTI**papers

Objective

To understand the language of metric prefixes. To learn how to make metric conversions by moving the decimal point.

Introduction

Converting metric units, one into another, is as easy as moving the decimal point. The difficult part is knowing which way to move it, and by how many decimal places. If your class doesn't already have a thorough understanding of decimals and metric prefixes (what class does?), this introduction is a necessity. Take all the time you need — a full class period or more:

First copy this figure on your blackboard. Make it large enough for all to see. (Notice that there are 5 zeros to the left of the one, and six zeros to the right.)

$$\overleftarrow{\boxed{DIVIDE}} \quad \overrightarrow{\boxed{MULTIPLY}}$$

$$000001{,}000000$$

ONE'S PLACE

Using a pointing stick or ruler, start at the one's place and multiply by 10's to the right. Introduce each new decimal place as you go: one, ten, hundred, thousand, ten thousand, hundred thousand, million. Begin at the one's place again and divide by 10's to the left: one, tenth, hundredth, thousandth, ten thousandth, hundred thousandth, millionth. Ask your class to repeat these decimal positions aloud while you silently point to each one. Move the pointer in descending and ascending orders at first. As your class becomes more familiar with these 12 decimal positions, skip randomly from one place to another.

Next, review how moving the decimal point multiplies or divides numbers by tens. Ask each student to copy your blackboard diagram onto a piece of scratch paper. Tell them to move their pencil points from the one's place (home base) to any new position you specify. You might ask them to divide 1 by 1000, for example.

$$00000{,}1000000$$

ONE'S PLACE

$$1/1000 = .001$$

After everyone agrees where the pencil point should end up, return to the one's place and try a new problem: multiply 1 by 1,000,000; divide 1 by 100; etc. Stress that moving the decimal *right* always *multiplies* by 10, 100, 1,000 . . . Moving the decimal *left* always *divides* by 10, 100, 1,000 . . .

Finally, introduce your class to the language of metrics. You state the prefix (kilo, hecto, deka . . .) and ask your class to chorus back the number (1,000, 100, 10 . . .). Or you state the number (.001, .01, .1 . . .) and ask your class to chorus back the prefix (milli, centi, deci . . .).

Once your students have thoroughly reviewed decimals, they are ready to begin this worksheet.

Lesson Notes

a. Students will complete these first 6 equations without difficulty. Before they proceed to part b however, ask them to review each equation they've written to see if it looks balanced. To maintain equality, the larger coefficient always goes with the smaller metric unit, and the smaller coefficient always goes with the larger. Hence, if

n(large unit) = m(small unit),

then, n *must* be smaller than m for the equation to balance.

b-f. You can think of these papers as forming a kind of staircase. To go *down* the stairs and stay equal, move the decimal to the *right* (multiply) in the same direction you descend. Descending from 1 paper yields these *equal* expressions:

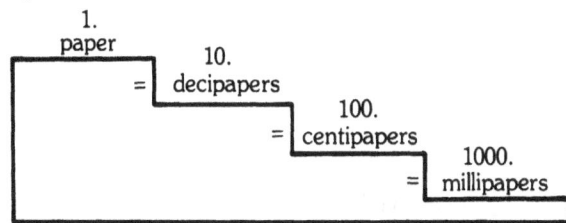

In like manner, to go *up* the stairs and stay equal, move the decimal to the *left* (divide) in the same direction you climb. Ascending from 5 papers yields these *equal* expressions:

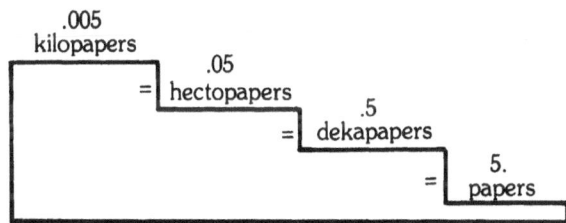

Apply this staircase concept to several equations in box b as examples, then let students complete all remaining equations on their own. Remind your class to always check each equation for balance by asking this important question: Is the *largest* metric unit multiplied by the *smallest* coefficient? If not, the equation is wrong.

If too many students still need too much help, complete this activity together as a class exercise. Work through each problem on the blackboard as your students fill in their own worksheets.

Extension

Do your students need more decimal practice? See DECIMAL FLOW CHART on page F-2 that follows.

Check Point

a.	b.	c.	d.	e.	f.
1000	10	10	.1	10	500
100	.1	1,000	.001	.1	5,000
.1	100	100,000	.00001	1	2,000
10	.01	100	.01	100	200
.001	1000	10,000	.0001	1	1
.01	.001	1,000,000	.000001	.01	15

NAME: CLASS:

Talking Metric **F-2**

LAYER UPON LAYER

1 Cut off 7 strips of ruled paper along the lines.

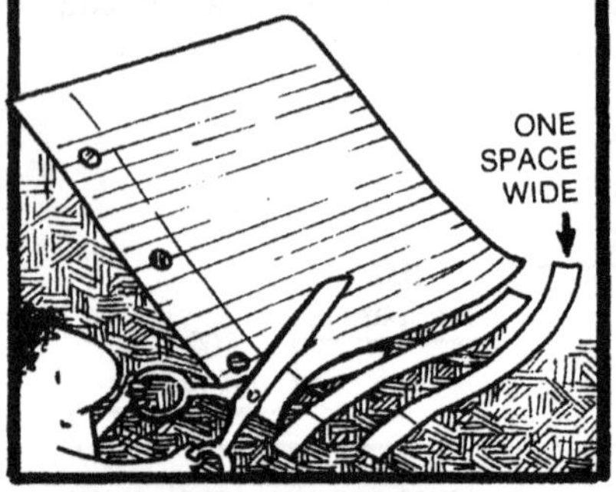

ONE
SPACE
WIDE

2 Fold one strip in half 4 times.

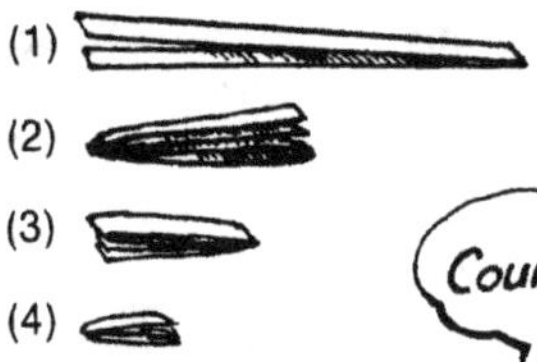

(1)
(2)
(3)
(4)

This unfolds into 16 parts.

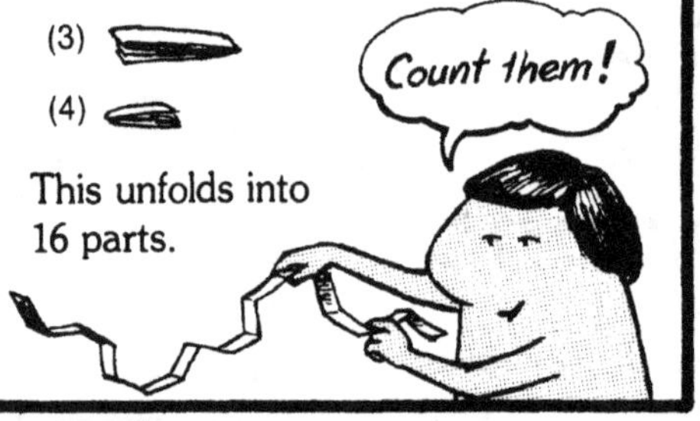

Count them!

3 Cut off 10 parts.

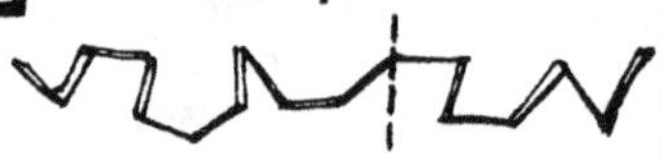

Fold into layers like this:

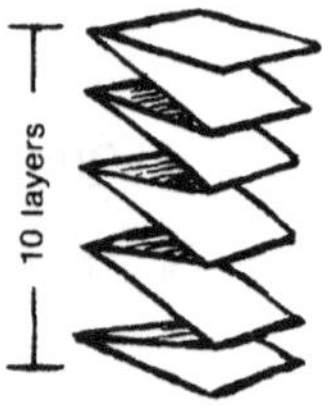

10 layers

4 Open a clothespin exactly 1 *DEKAlayer* wide.

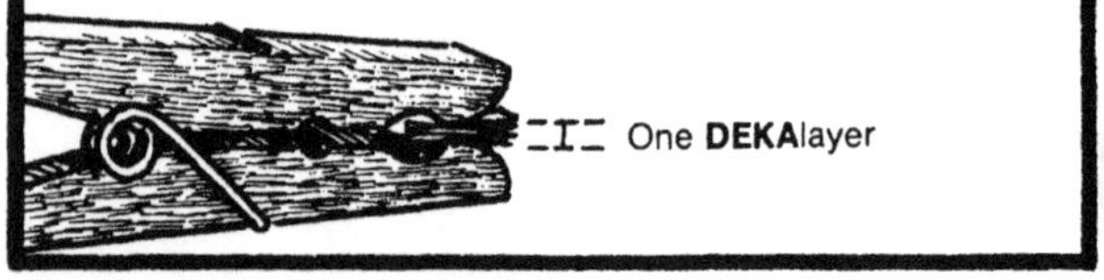

One **DEKA**layer

5 Add more layers of paper until it opens 1 *HECTOlayer* wide.

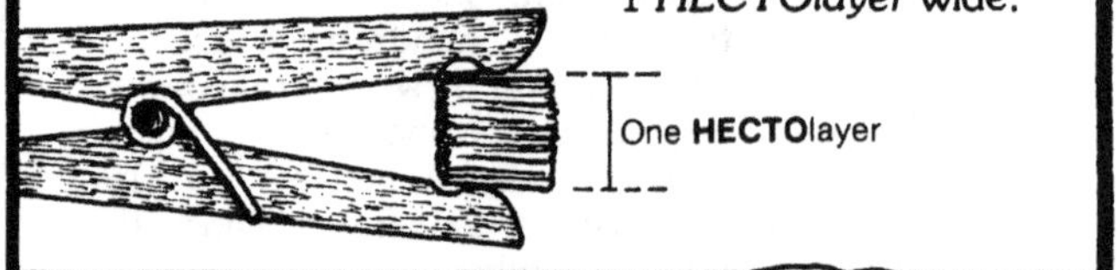

One **HECTO**layer

6 Draw a line to show how long 1 *KILOlayer* must be.

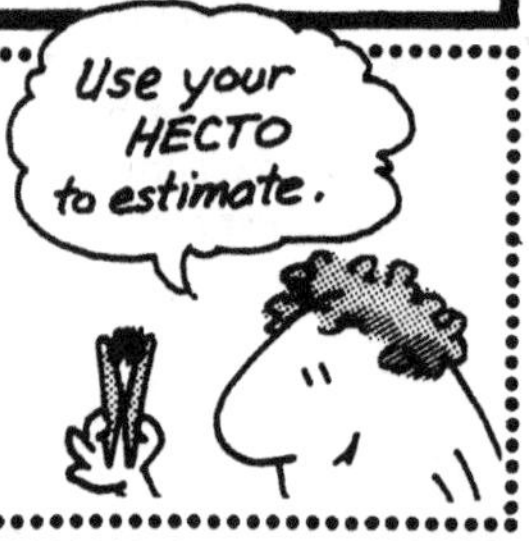

Use your HECTO to estimate.

7 Estimate how many sheets of ruled paper you can stack from your floor to your ceiling. Tell how you did this.

?

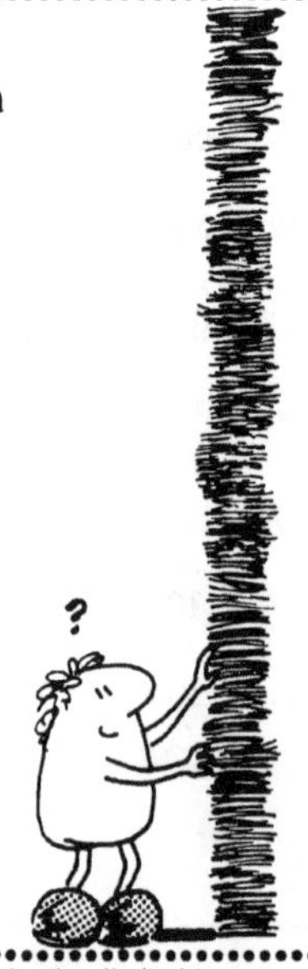

8 How tall must a warehouse be to hold paper stacked 1 million layers high? Show your math.

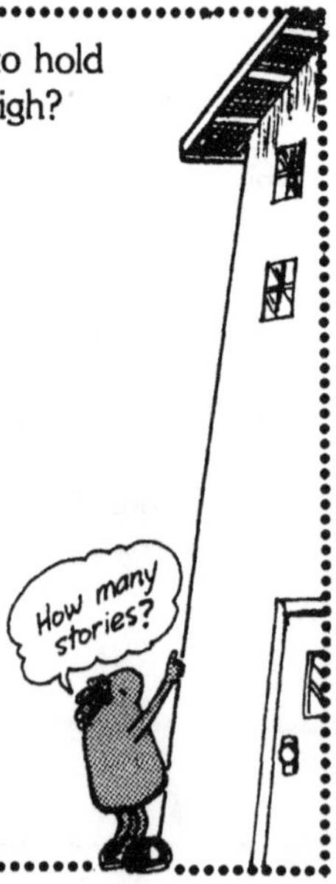

How many stories?

Objective

To estimate large numbers. To understand metric multiples in concrete terms.

Introduction

None required.

Lesson Notes

4-5. To make 10 layers (a dekalayer) and then 100 layers (a hectolayer), it's much easier to fold strips than to cut each piece individually.

Less coordinated students may need extra help in this step. 100 layers tend to "squirt" out of the clothespin jaws unless placed firmly inside.

DECIMAL FLOW CHART

Copy this flow chart on your blackboard so students can copy it onto their own papers.

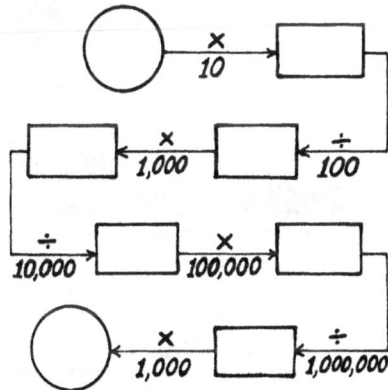

Ask students to input "1" in the top circle, then process it through the flow chart. If they perform each decimal operation correctly, they'll end up with the *same* number as an output. This makes the flow chart self-checking. Ask student to process other numbers through the flow chart so that input = output.

Finally ask students to invent their own decimal flow charts. Challenge them to make input = output using an entirely different sequence of decimal operations.

Check Point

6. Laying the hectolayer end to end ten times yields a line about this long:

A thousand sheets of relatively thin notebook paper reach about 8 cm. Your students may draw lines longer than this because they fail to compensate for the folds in the paper that flare out on either side.

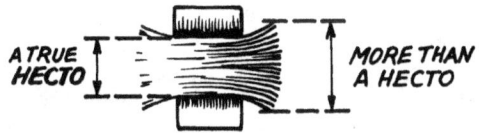

7. Here is one estimate for a ceiling 3 meters high. Student answers, of course, will vary considerably depending on the height of your particular room and their estimated length of 1 kilolayer.

3 meters x 100 cm/meter x 1,000 layers/8 cm
= 37,500 layers

Allow each student to follow an independent problem solving strategy. Some may cut a strip of paper to the length of 1 kilolayer, then use this to measure up the wall. Mathematicians might measure the height of the wall, then divide by the length of 1 kilolayer.

8. If 3 meters make 1 story, and each story accommodates 37,500 layers, then:

1,000,000 layers x 1 story/37,500 layers = 27 stories
(approx.)

KILOMETER RULER

I Find 7 units of measure on this ruler. Write them in order here:

LONGER
....
....
each unit is ten times
....
....
....
SHORTER

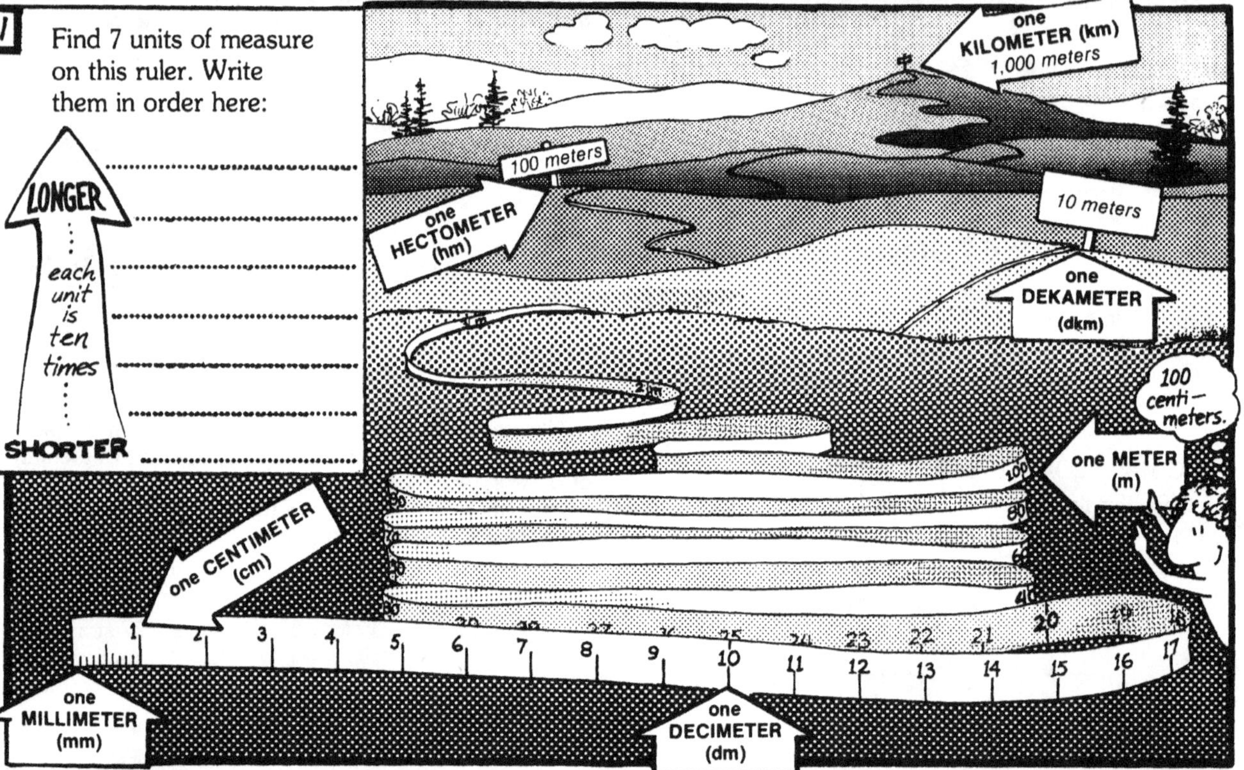

one KILOMETER (km)
1,000 meters

100 meters

one HECTOMETER (hm)

10 meters

one DEKAMETER (dkm)

100 centi-meters.

one METER (m)

one CENTIMETER (cm)

one MILLIMETER (mm)

one DECIMETER (dm)

2

a. How many *millimeters* in. . .	**b.** How many *centimeters* in. . .	**c.** How many *meters* in. . .	**d.** How many *kilometers* in. . .
1 cm?	50 mm?	100 cm?	1000 m?
5 cm?	35 mm?	50 cm?	10,000 m?
2.5 cm?	5 mm?	20 cm?	5,500 m?
7.8 cm?	1 mm?	19 cm?	5,597 m?
70 cm?	250 mm?	10 cm?	100 m?
1 dm?	1 dm?	3 cm?	99 m?
3 dm?	9 dm?	1 cm?	1 m?
1.6 dm?	.5 dm?	5 mm?	100 cm?
1 m?	1.2 dm?	1 mm?	67 cm?
4 m?	1 m?	300 mm?	1 cm?
1 dkm?	1 dkm?	1 km?	.8 cm?
1 km?	1 hm?	5 km?	1 mm?

Objective

To visualize how metric units fit together as multiples of 10. To practice expressing one measure in terms of another.

Introduction

This ruler stretches 1 full kilometer across your paper and imagination. All 7 units are clearly shown along the way. In a single glance you can appreciate the logic and structure of the whole metric system.

Such a remarkable ruler deserves special study. Consider leading your class on a pencil-point tour. As tour guide, you call out numbers on the ruler while your class points to each position you call with their pencil points.

Begin by counting the smallest millimeter divisions. The first centimeter division is easy: 1 mm, 2 mm, . . . 10 mm. Students will have to visualize how mm units divide the next cm: 11 mm, 12 mm, . . . 20 mm. Continue counting by larger and larger intervals: 25 mm, 30 mm, 50 mm, 100 mm, 200 mm, 500 mm, 1,000 mm. Past 1 meter stop only at clearly marked divisions along the way: 3 meters (3,000 mm), 1 dekameter (10,000 mm), 1 hectometer (100,000 mm), 1 kilometer (1,000,000 mm).

Next, take a tour of centimeters: 1 cm, 2 cm, 10 cm, 17 cm, 80 cm, 100 cm, 200 cm, 300 cm, 1,000 cm, 10,000 cm, 100,000 cm. Then tour meters and kilometers. Avoid measures that are not well represented in the drawing. Four meters, for example, hides behind the first hill.

As you circulate aboout your room, you'll easily recognize those who are pointing their pencils to the correct places, and those who need special help.

Lesson Notes

2. Here are 3 possible strategies for finding decimal equivalents:

(Method 1) Physically touch each given measure on the kilometer ruler. (From this point back to zero defines the ruler segment you must consider.) Now think about the size of the unit you are counting. (This is stated at the top of each column.) Ask how many of these units fit end to end into the ruler segment you have just defined.

(Method 2) Compare the problem you are working on with one you have just solved. If they contains identical units, then repeat the same pattern: If 2.5 cm equals 25 mm, for example, then 7.8 cm must equal 78 mm because it, too, is 10 times longer.

(Method 3) Solve any problem using a metric-stairs strategy similar to activity F-1. Draw this diagram on your blackboard for easy student reference.

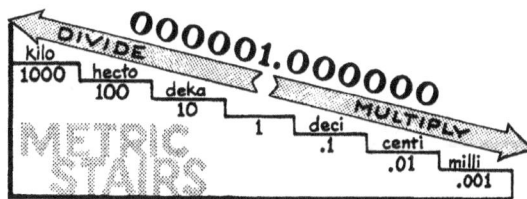

Check Point

1. kilometer
hectometer
dekameter
meter
decimeter
centimeter
milimeter

2. Don't expect too many to score a perfect 48 out of 48. Simply circle wrong answers, then ask students to try again. Some may achieve perfect scores the second time around. Others will need to return a third or fourth time. Don't count how many check points it takes to get all the answers right. Celebrate, instead, the final perfection.

a.
1 cm = 10 mm
5 cm = 50 mm
2.5 cm = 25 mm
7.8 cm = 78 mm
70 cm = 700 mm
1 dm = 100 mm
3 dm = 300 mm
1.6 dm = 160 mm
1m = 1,000 mm
4 m = 4,000 mm
1 dkm = 10,000 mm
1 km = 1,000,000 mm

b.
50 mm = 5 cm
35 mm = 3.5 cm
5 mm = .5 cm
1mm = .1 cm
250 mm = 25 cm
1 dm = 10 cm
9 dm = 90 cm
.5 dm = 5 cm
1.2 dm = 12 cm
1 m = 100 cm
1 dkm = 1,000 cm
1 hm = 10,000 cm

c.
100 cm = 1 m
50 cm = .5 m
20 cm = .2 m
19 cm = .19 m
10 cm = .1 m
3 cm = .03 m
1 cm = .01 m
5 mm = .005 m
1 mm = .001 m
300 mm = .3 m
1 km = 1,000 m
5 km = 5,000 m

d.
1000 m = 1 km
10,000 m = 10 km
5,500 m = 5.5 km
5,597 m = 5.597 km
100 m = .1 km
99 m = .099 km
1 m = .001 km
100 cm = .001 km
67 cm = .00067 km
1 cm = .00001 km
.8 cm = .000008 km
1 mm = .000001 km

NAME: CLASS:

METRIC SQUARES

1 Get a sheet of METRIC SQUARES. Cut out all 42 squares *and* 3 labels.

2 Sort your squares into 3 labeled piles:

VOLUME — 6 SQUARES   **MASS** — 12 SQUARES   **LENGTH** — 24 SQUARES

3 Sort each pile into triplets — groups of 1 white, 1 grey, and 1 black that are equal.

VOLUME (2 triplets)   **MASS** (4 triplets)   **LENGTH** (8 triplets)

TRIPLETS are always EQUAL

Record each set below!

4 Write each triplet in the correct space below.

VOLUME

One liter is ..... *As much as* .... and equals...........

One milliliter is .... and equals...........

MASS

One kilogram is .... and equals...........

One gram is .... and equals...........

One gram is .... and equals...........

One milligram is .... and equals...........

LENGTH

One kilometer is .... and equals...........

One meter is .... and equals...........

One meter is .... and equals...........

One meter is .... and equals...........

One centimeter is .... and equals...........

One centimeter is .... and equals...........

One millimeter is .... and equals...........

One millimeter is .... and equals...........

SAVE your metric squares!

Objective

To become familiar with 42 important interrelated facts about metric volume, mass, and length.

Introduction

In this activity students sort and match important metric quantities that are printed on squares of paper. These squares leave behind unusual metric units (hectograms, dekameters, and deciliters) in favor of more common fare. Examine a sheet of these squares from a student cut-out booklet or the reference page at the back of this book. You'll find kilograms, centimeters, milliliters and more, units that people use every day, important relationships that students should know.

To integrate these important metric facts into classroom language and thought, organize an old-fashioned flash card drill. Enlarge all 28 gray and black squares to serve as flash cards. They ask 28 important questions. Print each one in large, bold letters on a quarter-sheet of standard-sized paper. The white squares supply 14 possible answers. Write the correct response on the back of each flash card to serve as an answer key.

SAMPLE FLASH CARDS:

Drill your students with these flash cards now, and as often as necessary, until all the familiar forms of meters, liters and grams are instantly recognized as old friends.

Lesson Notes

1. Sheets of Metric Squares are available as student cutouts. You can also reproduce them, if you wish, from the line master in back of this book.

2. This step must be completed accurately. Otherwise, each pile will not properly subsort into equal triplets in step 3. The volume pile contains just six squares with 2 grays, reading "as *much* as . . ." The mass pile contains 12 squares with 4 grays, reading "as *heavy* as . . ."

3-4. Step 3 requires plenty of free desk space or an open floor. Sort the volume squares first. This is easy to do, because there are only 2 triplets. Next, sort the 12 mass squares into 4 triplets. Finally, sort the 24 length squares into 8 triplets. It's easiest to lay down all the white squares first. Then pair up the grays, and finally match the blacks.

Record results in step 4 as you go, or after matching all 14 triplets.

Remind students to save their metric squares for use in later card game activities.

Check Point

(Black or grey squares that are equivalent may be interchanged.)
VOLUME
One liter / as much as 4 glasses of water / 1000 ml.
One milliliter / as much as 15 drops of water / .001 l.
MASS
One kilogram / heavy as a chicken / 1000 g.
One gram / heavy as 2 paper clips / .001 kg.
One gram / heavy as 1/2 bottle cap / 1000 mg.
One milligram / heavy as a bread crumb / .001 g.
LENGTH
One kilometer / long as a 10 minute walk / 1000 m.
One meter / wide as doorway / 1000 mm.
One meter / long as your leg / 100 cm.
One meter / long as a monkey's tail / .001 km.
One centimeter / thick as blackboard chalk / 10 mm.
One centimeter / long as a fingernail / .01 m.
One millimeter / wide as a pinhead / .1 cm.
One millimeter / thin as a coin / .001 m.

FACE-UP

1 Make a grid to fit your METRIC SQUARES.

Like tic·tac·toe!

2 Shuffle your deck of 42 cards to mix them well.

3 Place the top 10 cards *face up* on your grid.

Start with 2 cards in the middle...

...1 card everywhere else.

4 Search for 2 *equal* cards among the 9 on your grid...

Hmm — 2 equal cards?

DRAW PILE (face up)

...Cover each pair you find with 2 new cards from your draw pile. Put them *face up* so they can form new pairs.

Equal!

5 Keep playing until...

...you *win!* (all cards played)

Matched them all!

DRAW PILE GONE

...you *lose.*

(cards left in the draw pile)

...stuck...

Teacher check

6 Now try these games:

SOLITARY FACE-UP

I see a pair!

Repeat steps 2-5 on your own.

TEACHER ✔: ☐ ☐ ☐

COOPERATIVE FACE-UP

your turn...

...1 mg is as heavy as a bread crumb!

Take turns finding 2 equal cards. Say each pair out loud.

TEACHER ✔: ☐ ☐ ☐

COMPETITIVE FACE-UP

100 cm = 1 meter!

Oooh — now we're TIED!

Try to be first to call out pairs. Keep score.

TEACHER ✔: ☐ ☐ ☐

SAVE your squares.

Objective

To firmly link metric units with common conversion factors and concrete images.

Introduction

None required.

Lesson Notes

1. Lay out this tic-tac-toe grid with masking tape, if you have an abundant supply. Otherwise tell students to use blackboard chalk.

3. There are 42 cards in the deck. Placing an extra in the middle leaves an even 32 in your hand, instead of an odd 33. This means you can discard pairs until you're left with none, instead of one.

4. The object is to match equal *pairs*. Remembering the last activity, some students may still think in terms of triplets.

5. There are 9 categories of metric cards. Each category contains from 3 to 9 matching cards. Here's the breakdown:

6	millimeter
6	centimeter
9	meter
3	kilometer
3	milligram
6	gram
3	kilogram
3	milliliter
3	liter
42	metric squares

If you can recognize pairs, your chances of winning are excellent. Generally, 1 to 3 pairs are available for matching at all times. However, you can occasionally lose. This happens when all 9 categories turn up — a different one in each of the tic-tac-toe positions. In our experience, this kind of no-match board occurs in about 1 out of 5 games. (Perhaps someone can calculate for us the theoretical probability of this happening.)

It's not easy to find a genuine no-match board. To confirm that no pairs exist, you have to mentally eliminate 36 different pairing combinations. Usually a pair is hiding somewhere on the board that you simply overlooked.

A true no-match board has one (and only one) card from each of the 9 categories. You can most easily verify that there are no pairs by taking this simple inventory: check off each tic-tac-toe position against this list of 9. If just one category is missing, then there *must* be a match hiding somewhere on your board.

In practice, beginners make many errors. They don't see pairs that really are there. They do see pairs that aren't — an easy way to win the game. Whether they win or lose is irrelevant. It's how they play the game. Those who learn are playing just as they should.

The teacher check box in step 5 implies that students should check with you before proceeding. Make sure they understand the fundamentals of Face-Up, and have played at least 1 complete game.

6. Some students thrive on the excitement of competition; others abhor it. Some work cooperatively together; others work best alone. In this step you can have it your way: play all three games, or concentrate just on the ones you like best. Remind your class to stay on task until they collect the number of game-completion teacher-checks you assign. We suggest that you require three.

Notice that both Cooperative and Competitive Face-Up ask students to call out each discovered pair aloud. This is important for several reasons. First, learning is enhanced with audio feed-back. Second, students may recognize numbers without being able to read them aloud. They need to practice saying "one thousandth of a gram" when they see .001 g. Third, calling each pair out loud allows an opponent to double-check another's answer, resulting in fewer game errors.

Check Point

Have metric recognition skills in your class improved as a result of playing Face-Up? You can assess progress by drilling once again with the flash cards you made in the last activity. Recognition times should be shrinking below one second by now. Instantaneous recognition is your goal.

NAME: _____   CLASS: _____

METRIC RUMMY

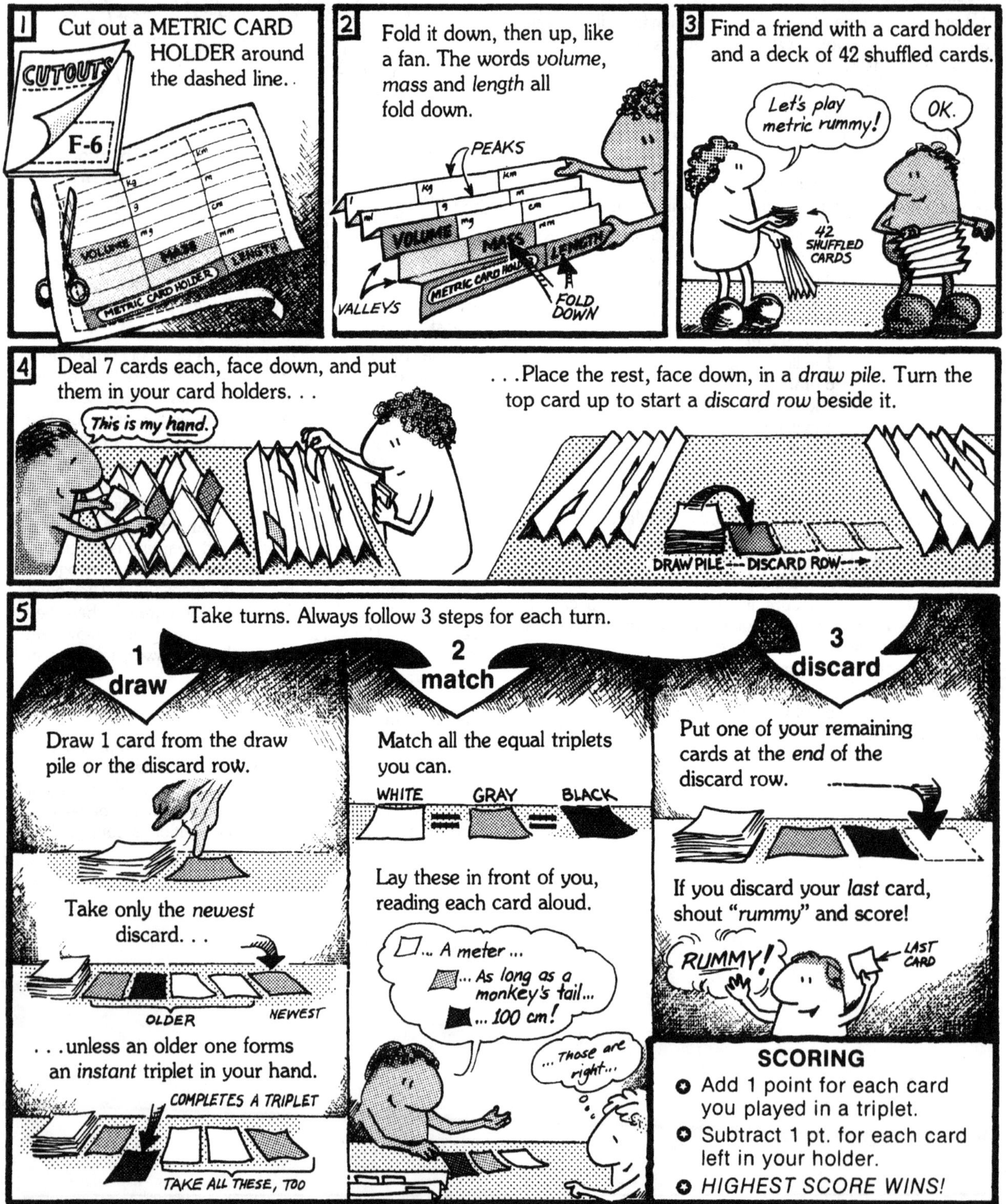

1 Cut out a METRIC CARD HOLDER around the dashed line.

2 Fold it down, then up, like a fan. The words *volume*, *mass* and *length* all fold down.

PEAKS

VALLEYS

FOLD DOWN

3 Find a friend with a card holder and a deck of 42 shuffled cards.

Let's play metric rummy!

OK.

42 SHUFFLED CARDS

4 Deal 7 cards each, face down, and put them in your card holders...

This is my hand.

...Place the rest, face down, in a *draw pile*. Turn the top card up to start a *discard row* beside it.

DRAW PILE ← DISCARD ROW →

5 Take turns. Always follow 3 steps for each turn.

1 draw

Draw 1 card from the draw pile *or* the discard row.

Take only the *newest* discard...

OLDER   NEWEST

...unless an older one forms an *instant* triplet in your hand.

COMPLETES A TRIPLET

TAKE ALL THESE, TOO

2 match

Match all the equal triplets you can.

WHITE   GRAY   BLACK

Lay these in front of you, reading each card aloud.

.. A meter ...
... As long as a monkey's tail...
... 100 cm!

...Those are right...

3 discard

Put one of your remaining cards at the *end* of the discard row.

If you discard your *last* card, shout "rummy" and score!

RUMMY!

LAST CARD

SCORING
- Add 1 point for each card you played in a triplet.
- Subtract 1 pt. for each card left in your holder.
- *HIGHEST SCORE WINS!*

Objective

To memorize metric relationships in a fun way.

Introduction

When was the last time you learned to play a card game by reading a book of Hoyle? Chances are, you learned most of the games you know by looking over someone else's shoulder. While it is possible for your class to learn Metric Rummy by studying this worksheet (we think the rules are very clear), a demonstration game will help your students learn much faster.

Lesson Notes

1. Metric card holders are available as student cutouts. You can also reproduce them, if you wish, from the line master in back of this book.

2. Those who fold the first flap up instead of down are starting wrong. This will throw the whole fold sequence out of step, reversing all the peaks and valleys.

3. Any number of students can play Metric Rummy. So that players can take turns more frequently, we recommend two players per game, or three at most.

4. Each card fits into one of nine categories in the card holder. Once classified in the correct position, you can identify triplets at a glance. Unfortunately, your opponent can infer what cards are in your hand as well, by astutely observing where they are positioned. Advanced players should disguise their hands by intentionally placing certain cards in the wrong category.

5. DRAW. Beginners tend to overlook potential triplets in the discard row, so it grows longer than it should. Stress how easy it is to capture discards. You only need to form 1 instant triplet from all the cards you already hold *plus* all discards on the table.

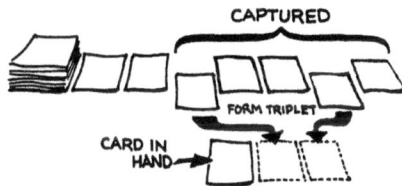

It is only fair to first show your opponent how you'll form the triplet, before you scoop up your captured cards.

Some games last long enough to empty the draw pile. If this happens, continue play without it. Unless you reach back to match an instant triplet, this means you'll always have to draw the last discard. As long as all players in the game don't draw then discard the *same* card, the cards should continue to circulate and lead to an eventual winner. If a discard stalemate does occur, enforce this special no-draw-pile rule: the card you draw can't be discarded again until your next turn.

5. MATCH. Except for playing an instant triplet to capture old discards, you don't have to play your other triplets unless you want to. By holding them in your hand, it may be possible to play all your cards at once. Then you can rummy while your opponent still has many cards. However, should your opponent go out first, the cards you hold count against you. It's dangerous to hold your triplets too long.

As you lay out a triplet on the table, be sure to read each card aloud. This allows your opponent to verify an accurate match. This is important. Any mismatched triplet, played in error now, creates other cards in the deck that have no match at all. Double-checking helps insure that the game will play out to an eventual rummy.

5. DISCARD. Traditionally, your last card must be discarded to rummy; it can't be played as a matched triplet. Beginners will play the last card any way they can, according to Hoyle or not. You can introduce this finer rule later, if you wish, or decide to ignore it completely.

In real rummy, you play a match of several games until the winner receives some arbitrary cumulative total. Here, we've opted for a shorter version: the highest score wins. Try this short version first. Play to a mutually agreed-upon total if enthusiasm remains high.   (A score of "30" requires 2, sometimes 3 games to reach.)

Check Point

Have metric recognition skills in your class improved as a result of playing Metric Rummy? Make a final assessment of progress by drilling once again with the flash cards you made previously. If recognition times are under 1 second, if your students can fluently express themselves using metric facts, congratulations! You have taught something worth knowing.

G. SIGNIFICANT FIGURES

G-1 the last digit

G-2 certain / uncertain

G-3 dial-a-measure

G-4 line measure

G-5 meter measure

G-6 body measure

ACCURATE MEASURE = ALL FIGURES YOU KNOW for CERTAIN ... + 1

Estimated, therefore UNCERTAIN

No physical measurement can be known with absolute accuracy. The best any scientist can do is to make sure that all digits in any measurement are significant.

ruler

This stick, for example, is correctly measured to 3 significant figures. It is 4.53 units long. The 4 and 5 are certain, the 3 is estimated. Neither 4.5 units nor 4.532 units are significant readings. The first measurement is

rounded off; the last digit in the second measurement is only a wild guess.

Significant figures stretch any measuring instrument (whether a crude meter stick or sophisticated vernier caliper) to its absolute limit of accuracy. That's the best anyone can do. Good science requires nothing less.

These activities provide your class with lots of measuring practice. No rounding off please. No extra fantasy figures that mean nothing. Students should write all the figures they are sure about, then estimate one more. Learn this now, and they will be reliable collectors of scientific data later.

——— EVALUATION ———

Each question evaluates a single activity from SIGNIFICANT FIGURES as numbered. Use any combination to frame a formal exam or an informal review: Copy these questions on your blackboard, construct your own ditto master, or photocopy the questions while masking out the rest of the page. Evaluate in ways that suit your own teaching style, enabling your students to learn and enjoy science.

Questions

G-1
Find the correct measure for each arrow.

G-2
Two students measure the length of the *same* index card. One finds its length to be 17.68 cm while the other gets 17.69 cm. Explain why these students get different answers.

G-3
Carlos and Sara each used one of the rulers below to correctly measure the length of a toothpick. Carlos measured 4.26 cm while Sara measured 4.3 cm. Which ruler did each student use? Explain how you know.

G-4
a. Find the length of stick x and stick y. Be sure to estimate the last digit.

b. Which side of the ruler is more accurate? Why?

G-5
Get a paper clip and a coin. Measure each by laying them on this ruler.

Underline the uncertain part of each measure.

G-6
Two students walk the length of a field to measure its length. One finds its length to be 290 paces, while the other gets 313 paces. Explain why these students get different answers.

Answers

G-1
a = 12.4 cm
b = 12.9 cm
c = 14.0 cm
x = 15.35 cm
y = 16.07 cm
z = 16.98 cm

G-2
Each student has measured the index card accurately. All certain figures agree. The answers don't need to agree in the last digit because it is estimated, and thus uncertain.

G-3
Carlos used ruler B because he estimated between mm intervals to the nearest .01 cm. Sara used ruler A because she estimated between cm intervals to the nearest .1 cm.

G-4
a. x = 3.4 cm
   y = 4.48 cm
b. The bottom (side y) is more accurate because it is subdivided into smaller, millimeter intervals. This allows you to estimate between the spaces to the nearest tenth mm (.01 cm).

G-5
Answers will vary depending on the particular paper clip and coin you provide. Here is one result:
paper clip = 3.22 cm
coin = 1.97 cm

G-6
The student who measured out 313 paces has smaller steps than the student who measured out 290 paces. To get the same answer (within acceptable limits of error), they need to use a better standard for measure, a meter tape, for example.

SEQUENCING

SIGNIFICANT FIGURES teaches students how to measure accurately with centimeter rulers. It logically follows F, but can easily be used earlier or postponed until later.

Related Activities: D—E---F---**G**

MATERIALS

Here is everything your students will use for the next 6 activities on SIGNIFICANT FIGURES. Materials printed in normal type are part of the core 15-things-in-a-box inventory that support all 100 activities. Materials printed in *italics* are additional local materials that you provide or ask your students to bring from home. Pencil and paper are already assumed and therefore unlisted. Each item is numbered with the activity where it is first used.

(G-3) Scissors.
(G-3) *Medium-sized tin cans.*
(G-3) Clear tape.
(G-3) Thread.
(G-3) *Small coins.*

FURTHER STUDY

Use problems like these plus "extension" ideas in SIGNIFICANT FIGURES to lead your students beyond worksheet activity into original research and investigation. Each discovery leads to more questions, deeper questions, better questions than these. Answering them is what good science is all about.

Who first decided how long a meter should be? How is it defined today? Research the history of the meter and write a report.

Have inventors ever been able to improve upon the accuracy of a ruler? Write a report on the vernier caliper.

Resolved: that all countries should adopt the metric standard world wide. Do you agree or disagree? Hold a class debate.

NAME: CLASS:

THE LAST DIGIT

1

Write the correct measure in each box. *Always* make the last digit a *zero* when the arrow points directly to a line.

example:

30.60 cm

ZERO means ON THE LINE.

cm: 30 31 32 33 34 35 36 37 38 39 40 41 42 43 44

h. c.   i.   a.   b. e.   d. j.   f. g.

☐ TEACHER CHECK

a.	b.	c.	d.
e.	f.	g.	h.	i.	j.

2

These arrows point *between* lines. Imagine that each space is divided into 10 parts, then *estimate* which tenth comes closest to the arrow.

example:

60.3 cm

I estimate about 3/10 of the way between.

cm: 60 61 62 63 64 65 66 67 68 69 70 71 72 73 74

a.   g.   i.   f.   c.   d.   b. j.   e. h.

☐ TEACHER CHECK

a.	b.	c.	d.
e.	f.	g.	h.	i.	j.

3

Write each of these measures with 4 digits. Be careful about the last digit.

ESTIMATE if between the lines. Write ZERO if on the line.

TWEEN   ON

cm: 41 42 43 44 45 46 47 48 49 50 51 52 53

g. c.   h. f. d.   e. i. j.   b.   a.

☐ TEACHER CHECK

a.	b.	c.	d.
e.	f.	g.	h.	i.	j.

Objective

To learn how to read a ruler accurately, estimating the last digit.

Introduction

Carefully draw an *enlarged* version of this centimeter ruler on your blackboard. Make the divisions as wide as possible, and keep them uniform. The ruler should cross most of your board.

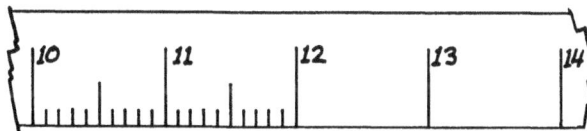

Count across this ruler, pointing out each division with a long pencil. Ask your class to join you in unison: 10.0, 10.1, 10.2, 10.3, . . . , 11.8, 11.9, 12.0, 13.0, 14.0. Point out the significance of attaching a zero to the to the last 3 numbers. This zero means your pencil pointer actually touches the division line, instead of someplace nearby.

Next point to places on the right end of the ruler *between* 12 and 14, challenging students to read each pencil position. Students should chorus back numbers with 3 digits like 12.5, 13.8, 13.1, etc. Explain that accurate measurements require writing down all the figures known for sure, then estimating between the lines to get the last digit.

Finally, point to places on the left end of the ruler between 10 and 12. Challenge students to again read back each pencil position as accurately as they can. All measurements *between* divisions must contain 4 significant figures, 3 that are certain plus the last that is estimated (10.53, 11.78, 11.11, etc.). All measurements *on* a division line also contain 4 significant figures, 3 that are certain plus the last "0" that is estimated to hit a division mark (10.20, 11.90, 11.10, etc.).

Notice that the accuracy of the ruler determines the number of significant figures. Three significant figures were possible when reading between 12 and 14. Four significant figures were possible when reading between 10 and 12 because the space was subdivided into smaller divisions.

Lesson Notes

1-3. Notice that each step has a teacher-check box. These steps conceptually build upon one another. As such, students should check their answers with you first (or with each other) before proceeding to the next step.

Forgetting to write units is a careless habit that students easily fall into. To break this habit, make a firm rule: on this worksheet and all others, each measure must express "how much" of "what". Half answers are not acceptable.

3. Arrow j may require a special explanation: The arrow points to the mm mark that's sandwiched between 49.9 and 50.1. There is no doubt that this division is 50.0. That is to say, 50.0 has 3 *certain* figures. Does the arrow hit square on the line (50.00)? Does it pull to the left (49.99)? Does it fade right (50.01)? Because the hundredth place is estimated, who can say for sure. Pending more sophisticated technology — a magnifying glass might do — one answer is as worthy as the next.

Check Point

All measurements in this activity must express *both* a number and a unit.

1. These answers have 4 significant figures. Allow no variations.
   a. 36.20 cm
   b. 37.50 cm
   c. 33.40 cm
   d. 40.10 cm
   e. 38.30 cm
   f. 42.70 cm
   g. 43.80 cm
   h. 32.90 cm
   i. 34.60 cm
   j. 41.00 cm

2. These answers have 3 significant figures. Allow variations of plus/minus .1 cm.
   a. 62.5 cm
   b. 70.2 cm
   c. 67.9 cm
   d. 69.1 cm
   e. 72.4 cm
   f. 66.6 cm
   g. 63.7 cm
   h. 73.3 cm
   i. 64.8 cm
   j. 71.0 cm

3. These answers have 4 significant figures. Allow variations of plus/minus .02 cm.
   a. 53.40 cm
   b. 51.53 cm
   c. 42.30 cm
   d. 45.90 cm
   e. 47.85 cm
   f. 44.77 cm
   g. 41.22 cm
   h. 43.59 cm
   i. 48.95 cm
   j. 50.00 cm

Metric Clock!

NAME: CLASS:

Significant Figures **G-2**

CERTAIN / UNCERTAIN

1 Write 2 possible measures for each arrow that points to RULER A . . .

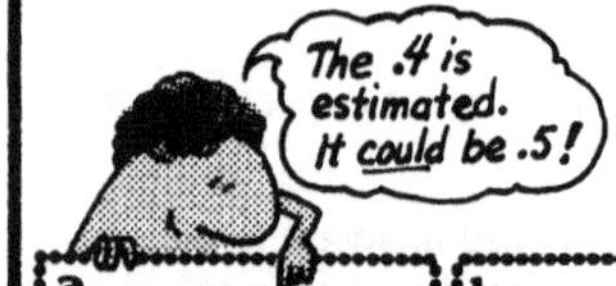

The .4 is estimated. It could be .5!

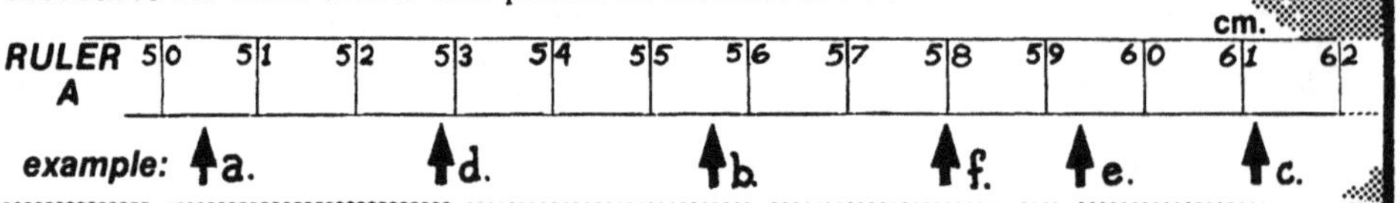

RULER A
cm.
50 51 52 53 54 55 56 57 58 59 60 61 62
example: a. d. b. f. e. c.

a.	b.	c.	d.	e.	f.
50.4 cm					
50.5 cm					

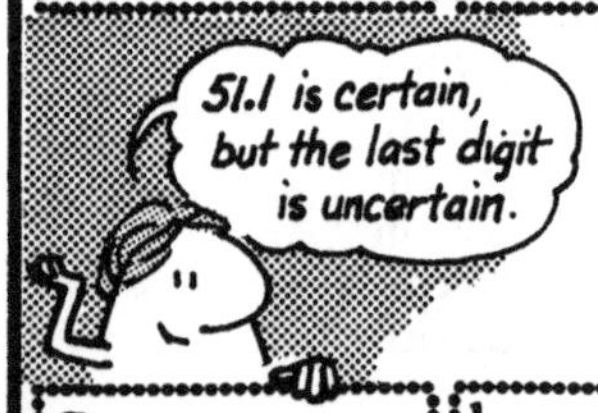

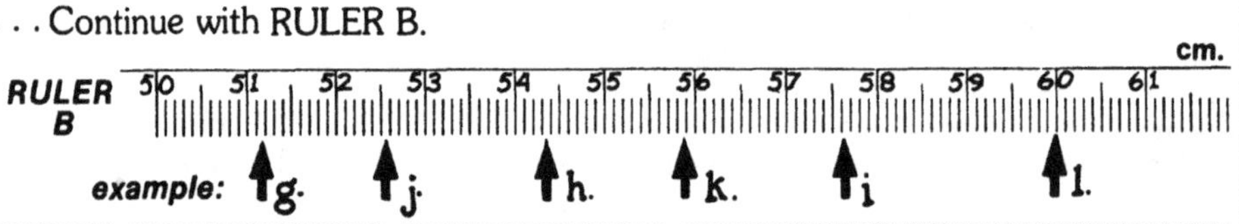

. . . Continue with RULER B.

51.1 is certain, but the last digit is uncertain.

RULER B
cm.
50 51 52 53 54 55 56 57 58 59 60 61
example: g. j. h. k. i. l.

g.	h.	i.	j.	k.	l.
51.17 cm					
51.16 cm					

2 For each measurement you just made. . .

. . . *underline* what is certain. . .

. . . *circle* what is uncertain.

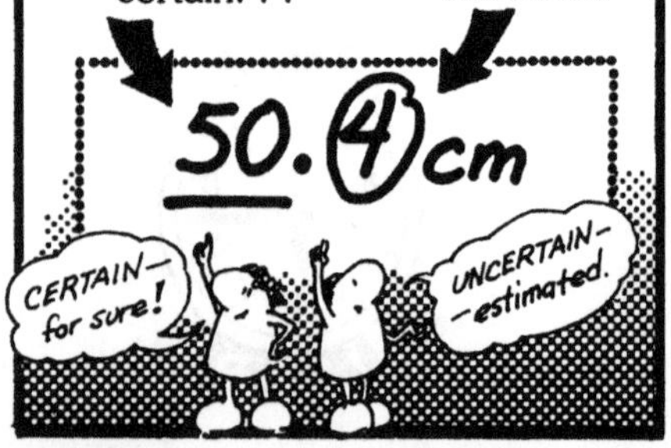

50.4 cm
CERTAIN— for sure!
UNCERTAIN— estimated.

3 Which ruler is more accurate, A or B? Why?

4 Can you make a ruler that has no uncertainty? Explain.

5 These folks are having an argument.

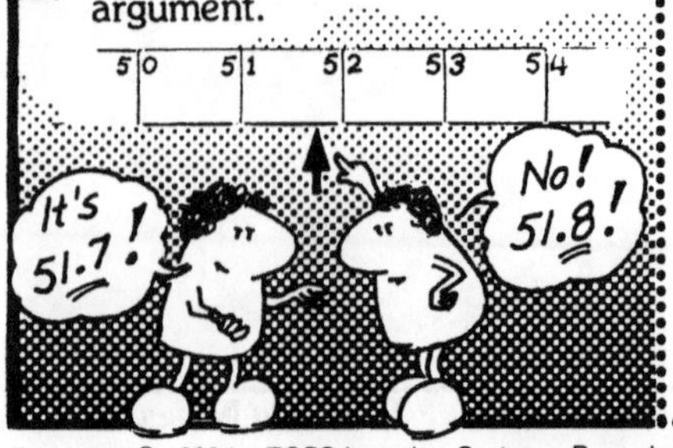

50 51 52 53 54
It's 51.7!
No! 51.8!

Can both be right? Explain.

Objective

To distinguish between certain figures and uncertain figures. To appreciate that no measurement is exact.

Introduction

None required.

Lesson Notes

1. To estimate between the lines on ruler A, you need to mentally divide the centimeter space into 10 parts, then decide which part comes closest to the arrow. It's not easy to decide, because each arrow is carefully positioned to fall *between* imaginary tenths.

Because ruler A is divided into centimeter divisions, uncertainty is properly expressed inside these large divisions to the nearest tenth cm. Ruler B, by contrast is divided into millimeter divisions. Uncertainty on this ruler is properly expressed inside these small divisions to the nearest hundredth cm.

Check Point

1. All measurements in this activity must express *both* a number and a unit.

RULER A	RULER B
These answers have 3 significant figures. Allow no variations.	These answers have 4 significant figures. Allow variations of plus/minus .02 cm.
a. 50.4 cm	g. 51.17 cm
50.5 cm	51.16 cm
b. 55.6 cm	h. 54.33 cm
55.7 cm	54.34 cm
c. 61.1 cm	i. 57.61 cm
61.2 cm	57.62 cm
d. 52.8 cm	j. 52.55 cm
52.9 cm	52.56 cm
e. 59.4 cm	k. 55.88 cm
59.3 cm	55.87 cm
f. 57.9 cm	l. 59.99 cm
58.0 cm	60.00 cm

2. The last digit in each measure is underlined because it is always estimated and therefore uncertain. If this uncertainty occurs between a nine and a zero, then other digits in the number are also affected. Consider arrow "l", the very last measurement. In this measurement all the numbers are uncertain because the arrow falls just a whisker to the left of the major 60 cm division. This doesn't imply that the uncertainty is any greater. It's still limited to the hundredths place. Rather, its proximity to a major division creates a kind of ripple effect across the entire number.

3. Ruler B is more accurate, because it is subdivided into smaller mm divisions.

4. No. There is always uncertainty between subdivided lines. No matter how closely these lines are placed, there is still uncertain space between them.

5. Yes. They only disagree over the last estimated digit. Because this digit is uncertain, both measurements are acceptable.

NAME: CLASS:

CUTOUTS
G-3

DIAL-A-MEASURE

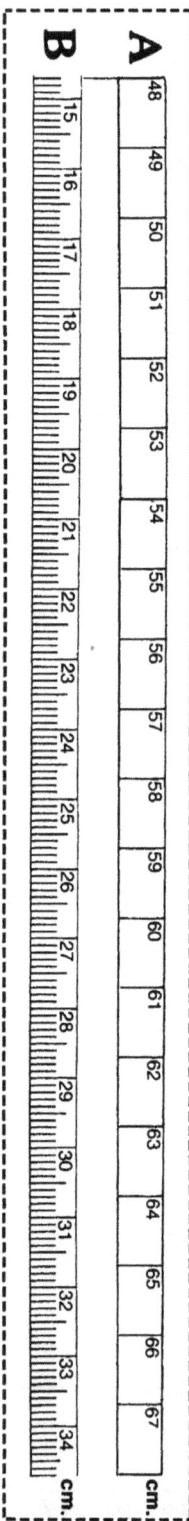

B **A**

(ruler B: 15–34 cm)
(ruler A: 48–67 cm)

1 Cut out the 2-ruler pattern along the outside dashed lines.

2 Wrap these rulers around an inverted can, up to the rim. Tighten and tape.

Tape only the ends.

3 Tape thread to the middle of the flat end. When you lift the can by its thread, it should hang straight.

CENTER

4 Tape a coin so it hangs just off the table.

Near the table.

5 Trim both ends.

6 Dial any chance measure. Then record and check your results.

This checks out! B+33.30=A

ALL UNITS ARE IN CENTIMETERS:

(a)
A 51.4
B 18.13
check:
18.13
+33.30
51.43

(b)
A
B
check:
+33.30

(c)
A
B
check:
+33.30

(d)
A
B
check:
+33.30

(e)
A
B
check:
+33.30

(f)
A
B
check:
+33.30

(g)
A
B
check:
+33.30

(h)
A
B
check:
+33.30

(i)
A
B
check:
+33.30

(j)
A
B
check:
+33.30

(k)
A
B
check:
+33.30

Objective

To practice reading a hairline (in significant figures) that crosses rulers calibrated in centimeters and millimeters.

Introduction

None required.

Lesson Notes

2. Tape only where the two ruler ends come together. By keeping the rest of the scale tape-free, there will be no rough edges to snag the thread as students dial measurements in step 6.

3. The point where the thread first meets the tape should be on center. This means the tape's *edge* (not the tape itself) will cross the middle of the can.

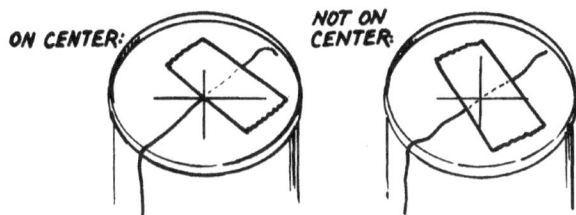

6. Use the can on a level surface. This keeps the thread in a vertical position, held with hairline accuracy by the weight of the coin. (Those with slanting desks may need to use the floor.) Never pick up the can to read the hairline. Rather, lower your head to get an eye-level view of the scale.

Students should randomly spin the thread around the can. After each spin they must record measures on rulers A and B in significant figures then confirm, as a self check, that the sum of measure B plus 33.30 yields a figure the rounds off to measure A.

Extension

Group your class by two's to play a cooperative measuring game called Agree/Disagree. Each pair of students needs just 1 dial-a-measure can. Follow these 3 simple rules using the A ruler *only*.

(1) Slide the hairline to any chance position.

(2) Both read the hairline in significant figures on the same ruler. Without talking to each other, write that measure on separate pieces of paper.

(3) Compare answers and score:

    ● Add 1 point if both measurements agree: uncertain digits can disagree by no more than 1.

    ● Subtract 2 points if both measurements disagree: certain digits disagree or uncertain digits disagree by more than one. (Don't subtract points below zero.)

    ● Both players win if they can reach a score of 5 points.

Once students win on the A ruler, challenge them to play again on the B ruler. This scale is more difficult because the estimated interval shrinks from a centimeter to only a millimeter. Parallax is now an important consideration. Moving the eye just a little off center to the right or left of the thread will produce readings that don't agree.

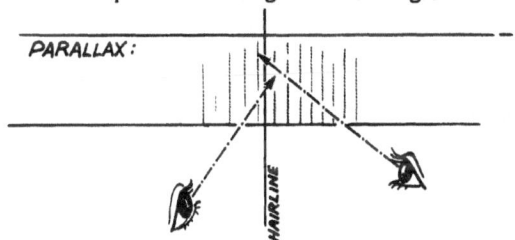

Check Point

All 10 answers are mathematically self checking. Spot check 1 or 2 to verify that the math is correct.

LINE MEASURE

1 Carefully cut out the *cm* ruler. Measure each line and write your answer in the box.

ESTIMATE THE LAST DIGIT— DON'T ROUND IT OFF!

2 Now cut out the ruler with *mm* divisions. Carefully re-measure each line and write your answer in the box.

WRITE UNITS ↓

WRITE UNITS ↓

a. 2.6 cm	**a.** _____	**a.** 2.65 cm
b.	**b.** __	**b.**
c.	**c.** _____	**c.**
d.	**d.** _____	**d.**
e.	**e.** _____	**e.**
f.	**f.** __	**f.**
g.	**g.** _____	**g.**
h.	**h.** ____	**h.**
i.	**i.** _	**i.**
j.	**j.** _____	**j.**

3 Which ruler is more accurate — the centimeter one or the millimeter one? Why?

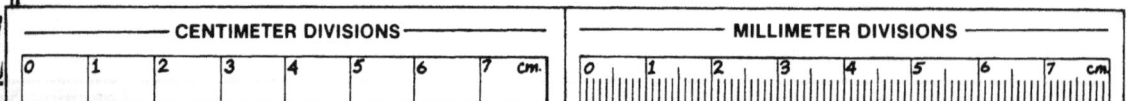

CUTOUTS

G-4

CENTIMETER DIVISIONS	MILLIMETER DIVISIONS
0 1 2 3 4 5 6 7 cm.	0 1 2 3 4 5 6 7 cm

Objective

To practice measuring accurately with a ruler. To recognize that estimating is always necessary, no matter how accurate the ruler.

Introduction

None required.

Lesson Notes

1-2. Many rulers, like these small cut-outs, have scales that are recessed away from the ends. Students who make the ruler's physical end their starting point, instead of the zero mark on the scale, will record lengths that are consistently too short.

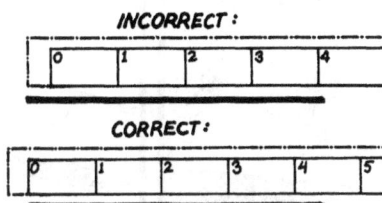

This end space is intentionally exaggerated to make it more obvious that the scale's end and the ruler's end are not congruent. Students should preserve this end space by cutting around the *outside* line of each ruler. While cutting, they must leave the bottom scale straight and even. If gaps appear, due to ragged cuts, the ruler can't be positioned close to a line without covering some parts of it.

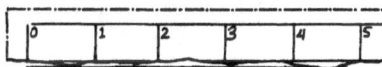

Sample answers are provided at the top of each column. Your students simply need to continue in the pattern of these examples — estimating centimeters to the nearest tenth on the left and hundredth on the right.

We all have a tendency to make things come out even — to shift the ruler a little this way or that so lines conveniently end on major divisions. Be alert for zero bias, especially in the right column of numbers.

Students might check the consistency of their own answers *before* bringing their worksheets to you for a check point. If they mentally round off each measurements on the right to the nearest tenth, it should be equal (within .1 cm) to each measurement on the left. Answers that fail this internal test should be remeasured.

Check Point

1. These answers have 2 significant figures. Allow variations of plus or minus .1 cm.
a. 2.6 cm
b. 1.3 cm
c. 5.1 cm
d. 4.3 cm
e. 7.7 cm
f. 0.9 cm
g. 5.3 cm
h. 2.0 cm
i. 0.4 cm
j. 6.6 cm

2. These answers have 3 significant figures. Allow variations of plus or minus .03 cm.
a. 2.65 cm
b. 1.27 cm
c. 5.10 cm
d. 4.32 cm
e. 7.75 cm
f. 0.90 cm
g. 5.31 cm
h. 2.05 cm
i. 0.39 cm
j. 6.59 cm

3. The millimeter ruler is more accurate, because it has smaller subdivisions.

METER MEASURE

CUT OUTS

G-5

LEAVE THIS TAB ON

1 Carefully cut around the outside of the shaded box.

2 Now cut your shaded box into 5 long strips. Tape these strips in order — 20, 40, 60, 80 — to make a meter tape.

LINE UP WITH THE STRAIGHT EDGE OF YOUR TABLE.

3 Write your name on the back of your tape.

4 Find the dimensions of each object in centimeters. Measure accurately — don't round off.

a. THIS WORKSHEET:

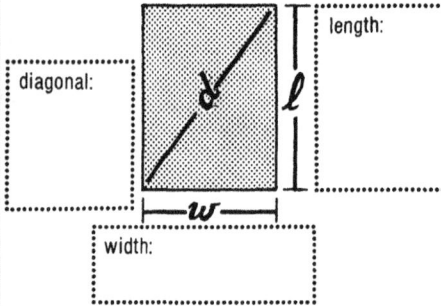

diagonal:

length:

width:

b. A TIN CAN:

circumference:

height:

diameter:

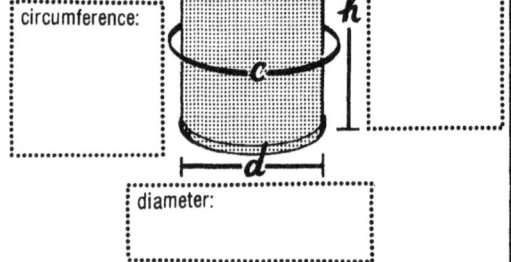

c. YOUR DESK TOP:

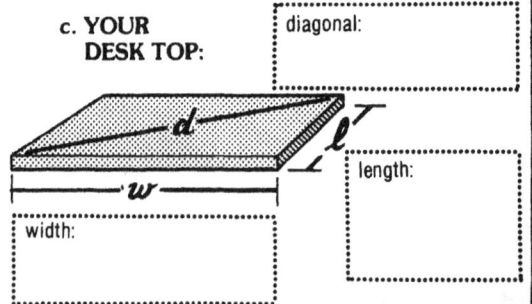

diagonal:

length:

width:

d. A COIN:

thickness:

diameter:

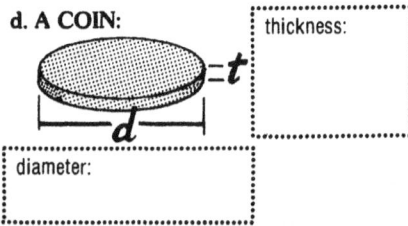

5 Underline the *certain* part of each measurement above. Check with a friend to make sure your *certain* figures agree.

CHECK!

Do *all* figures in each measure agree? Explain.

SAVE YOUR METER TAPE

Objective

To practice measuring physical objects with a meter tape. To estimate the last digit.

Introduction

None required.

Lesson Notes

1-2. To make an accurate meter tape, students must cut squarely on the lines, then tape the 5 strips together so the ends just touch. Avoid overlaps that make the ruler too short, and gaps that make the ruler too long. If it is well pieced together, the tape should measure close to a standard meter.

If you have photocopied this cutout instead of using the student cutouts, be aware that some copiers may significantly lengthen or shorten an image. Even slight photocopy distortions will add up to a significant cumulative error, over a distance of 1 meter.

4. The purpose of this exercise is to practice measuring accurately. Each measurement, therefore, must be estimated between the lines to the nearest tenth centimeter. Watch out for students who persist in rounding off to whole centimeters. Few, if any, of these measurements should come out even.

5. Remind students to save their meter tapes. They will use them again in the next activity.

Extension

Using measurements from step 4, students can verify two mathematical relationships that date back to the Greeks.
a. Use numbers from the rectangles you have measured to verify the Pythagorean theorem.

$$a^2 + b^2 = c^2$$

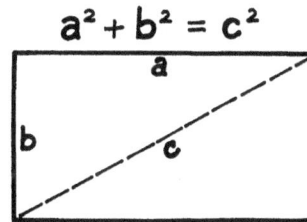

b. Use numbers from the cylinders you have measured to calculate pi — the ratio of the circumference to the diameter of any circle. ($\pi = 3.14$ approx.)

$$\pi = \frac{c}{d}$$

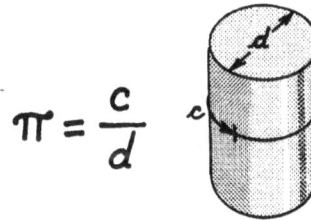

Check Point

4. Answers will vary, depending on what objects students measure.

5. Students should underline the certain figures in each measurements of step 4. Only these certain figures will agree. The last figures will sometimes disagree because they are estimated, and therefore uncertain.

BODY MEASURE

1   Use your meter tape to find how you size up: *Estimate* between lines, don't round off.

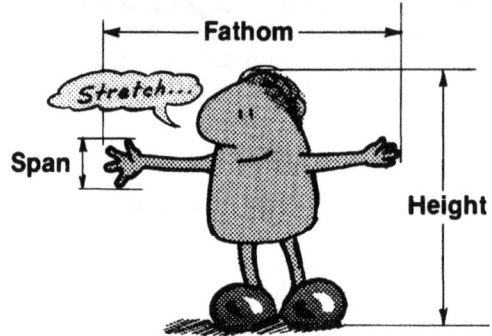

CIRCUMFERENCES:

Forehead: ....   Waist: ....

Neck: ....   Wrist: ....

LENGTHS:

Height: ....

Fathom: ....

Span: ....

2   A very famous scientist* found that 8 times your span equals both your height and your fathom:

8 spans = 1 fathom = 1 height.

Use math to show how closely *you* fit this rule.

*Leonardo da Vinci

3   Is it better to measure in spans or in meters? Explain.

Objective

To discover basic body proportions by making accurate measurements with a meter tape. To appreciate that body units of measure are not standard.

Introduction

Height is a dimension that is familiar to all. Spans and fathoms may require additional explanation. Ask your class to stand and stretch out their arms as far as possible: This is a fathom. Now, separate thumb from little finger as far as possible: This is a span. Emphasize "as far as possible." The mathematical relationships in step 2 work best if students stretch to full extensions.

Now ask a student with large hands to measure the length of your blackboard in spans. Follow up with another student who has small hands. Ask why the first measure is shorter than the second. (The blackboard is certainly not changing size.) Discuss the idea of a measuring standard — a unit that is the same for everyone.

Finally, propose that students measure the blackboard again, this time in fathoms. Ask your class to *predict* who would likely get the longest length (the smallest student); the shortest length (the biggest student).

Lesson Notes

1. Students will need to work in pairs so they can help each other hold the meter tape and read the scale. Each measure, of course, should be estimated between centimeter intervals to the nearest tenth. Units must accompany each answer.

Heights, fathoms, and spans are easiest to measure with pencil marks against a wall. Designate a special measuring area in your room. Corners work nicely. Tape scratch paper to the walls if you need to protect paint finishes.

Check Point

1. Student physical sizes will vary. All measurement must be significant, accurate to the nearest .1 cm.

2. Answers will again vary. Here is one sample calculation:

$$span = 17.0 \text{ cm}$$
$$8 \text{ spans} = 8 \times 17.0 \text{ cm} = 136.0 \text{ cm}$$
$$fathom = 131.1 \text{ cm}$$
$$height = 135.2 \text{ cm}$$

The last 3 measurements are roughly equal.

3. It's better to measure in meters. A span changes with body size, but a meter is standard — the same for everyone.

H. BULBS AND BATTERIES

Bulbs, dry cells and a few common TOPS materials are all you need to begin a fascinating study of electricity. No wire is required. The usual twists and tangles are replaced instead by flat, flexible conducting ribbon: aluminum foil backed by tape.

Where are the important contact points on a bulb and battery? Do cells in series produce more light than cells in parallel or opposition? What kinds of materials are conductors of electricity and which ones insulate?

Questions like these plus many others require no introduction on your part, little explanation later on. Answers follow naturally within the context of curiosity and active exploration. Students simply ask "what will happen if . . .," then set about to find the answer.

EVALUATION

Each question evaluates a single activity from BULBS AND BATTERIES as numbered. Use any combination to frame a formal exam or an informal review. Copy these questions on your blackboard, construct your own ditto master, or photocopy the questions while masking out the rest of the page. Evaluate in ways that suit your own teaching style, enabling your students to learn and enjoy science.

Questions

H-1
Draw a way to light a bulb with a dry cell. Draw another way that doesn't work.

H-2
Connect each bulb and dry cell with lines to show how to light the bulb.

H-3
Predict if these bulbs will light. Give a reason for each prediction.

a.   b.

H-4
Identify those cells in each pair that make a bulb shine brighter. Circle one of the a's, one of the b's and one of the c's.

a1   b1   c1

a2   b2   c2

H-5
Number these 5 groups of cells by how bright they make the bulb shine. Write "1" in the blank next to the brightest, "2" for the next brightest, and so on.

a. _____
b. _____
c. _____
d. _____
e. _____

H-6
You have discovered a UFO (unidentified fallen object) near your home. You want to find out if it's a conductor or insulator of electricity. Use words and pictures to tell what you would do.

H-7 Fill in the blanks with the correct numbers.

1 2
3 4
5 6

a. A bulb lights with holes 1, 4 and 4, 5. Therefore, the bulb *must* light with holes _____ as well.

b. Hole 1 lights *only* with 2. Hole 2 therefore *cannot* light with holes _____.

Answers

H-1
Answers will vary. Here are two.
This works.   This doesn't work.

H-2

H-3
a. Prediction: No. The bottom contact point on the bulb is not connected to the cell.
b. Prediction: Yes. Both contact points on the cell are connected to both contact points on the bulb.

H-4 (a1,) b1, c1
         a2, (b2,) (c2)

H-5 a. 4
         b. 2
         c. 3
         d. 1
         e. 5

H-6
Connect a piece of the UFO to a bulb and cell, as illustrated. If the bulb shines, it's a conductor. If not, it's an insulator.

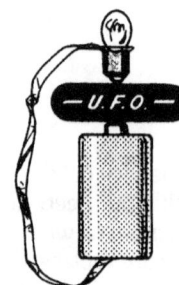

— U.F.O. —

H-7
a. 1 and 5. (Hole 1 is connected with 5 via 4.)
b. 3, 4, 5, or 6. (Holes 1 and 2 are interconnected. If hole 1 won't light with any other hole, neither can hole 2.)

SEQUENCING

BULBS AND BATTERIES is an introduction to simple electric circuits and how they work. Schedule it at any time, but follow up with I and J. **H** is a necessary prerequisite to both.

Related Activities: **H—I—J**

MATERIALS

Here is everything your students will use for the next 7 activities on BULBS AND BATTERIES. Materials printed in normal type are part of the core 15-things-in-a-box inventory that support all 100 activities. Materials printed in *italics* are additional local materials that you provide or ask your students to bring from home. Pencil and paper are already assumed and therefore unlisted. Each item is numbered with the activity where it is first used.

(H-1) Masking tape.
(H-1) Aluminum foil.
(H-1) Scissors.
(H-1) Size-D dry cells (1.5 volts).
(H-1) Flashlight bulbs. Use a size that is designed for 2 dry cells (3 volts).
(H-4) Paper clips.
(H-6) Rubber bands.
(H-6) *Small coins.*
(H-6) An assortment of metal and non-metal objects to test for conductivity. See "Preparation" in teaching notes H-6.
(H-7) *Scratch paper.*

FURTHER STUDY

Use problems like these plus "extension" ideas in BULBS AND BATTERIES to lead your students beyond worksheet activity into original research and investigation. Each discovery leads to more questions, deeper questions, better questions than these. Answering them is what good science is all about.

Dissect a dry cell. (Ask an adult to help you.) Write a report about what you find.

Write a report about Thomas Edison and the electric light bulb.

Read about static electricity. Select one of these topics for further study:

     lightning
     capacitors
     electroscopes

IT WORKS!

1 Stick 30 cm of masking tape to a narrow strip of aluminum foil.

30 cm is a little longer than this page.

2 Cut around the *inside edge* of the masking tape.

Remove all excess foil.

CUT ALL AROUND

3 Fold the ribbon along its length; foil side *out*, tape side *in*.

TAPE   FOIL

4 Crease the fold along the edge of your table.

5 Use your foil ribbon to light a bulb with a dry cell.

How to make it light?

6 Draw your results below using pictures like these.

BULB

DRY CELL   RIBBON

These work:	**a1.**	**b1.**	**c1.**
Draw 3 different ways you tried that WORKED!			
These don't work: **a2.**	**b2.**	**c2.**	
Also draw 3 different ways you tried that DIDN'T WORK!			

TAPE YOUR ~~BRIDGE~~ TO YOUR DRY CELL

Objective

To discover by trial and error how to light a bulb with a dry cell and ribbon.

Introduction

This activity is discovery science at its best — pure joy. Don't spoil the experience by explaining how things should work ahead of time.

Lesson Notes

1-2. Some students may cut the foil into strips *before* they apply the tape. This doesn't work. Students who attempt this will create a mess and have to start over.

If you wish to avoid having your students handle (or mishandle) long strips of sticky masking tape altogether, you might pretape this foil yourself, then ask your class to begin by trimming the excess foil in step 2.

You can substitute other kinds of pressure sensitive opaque tapes (wide packaging tape, for example) in place of masking tape. Whether you use wide tape or narrow, always cut the foil ribbon to centimeter widths.

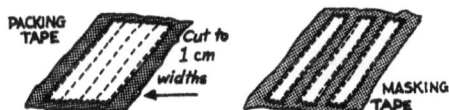

6. Students who have not had previous experience with bulbs and dry cells will need to do a lot of trial and error investigating. Such exploratory activity, of course, is just what you want to happen. Don't be too quick to come to the aid of puzzled students.

Very young students may become frustrated beyond their ability to cope. You can help them if you twist the foil ribbon around the top of the bulb collar. With this connection in place, they will easily discover the other connection and light the bulb.

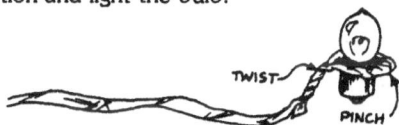

Some students may complain of being "shocked." Electricity produced by size-D cells is totally safe, not strong enough to harm anyone. What they actually feel is the *heat* generated when both poles of the cell are accidentally connected with foil ribbon. Without a light bulb properly connected in between to provide resistance, electrons flow

in sufficient quantity to noticeably heat the ribbon. Caution your students to immediately disconnect these "hot wires" to keep the cell from being drained of energy.

ENERGY-DRAINING HOT WIRE

It's conceptually important to clearly show how the contact points on the bulb and cell interconnect. Don't accept careless drawings where the bulb looks like a blob, the cell like a box, and the foil ribbon like the double lines of a railroad track. Ask students to redraw unclear pictures, using the simplified drawings of the bulb, dry cell and ribbon as models.

Remind students at the end of this activity to tape their names to their dry cells. Students who understand that they must continue to use their *own* dry cells throughout these activities, will more readily conserve energy and avoid energy-draining hot wires.

Check Point

6. These ways *work* (among others):

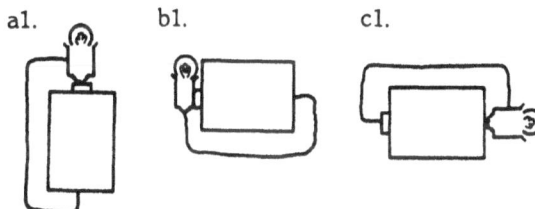

6. These ways *don't work* (among others):

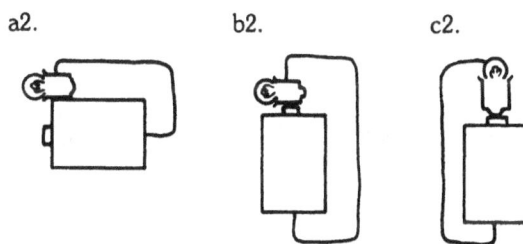

TO LIGHT OR NOT TO LIGHT

1

Let's work together!

Find a friend who also has a bulb and dry cell. Then solve each puzzle below and draw what you discover.

a. Light a bulb without touching it to the dry cell. (You may use 2 foil ribbons.)

b. Make the bulb shine extra bright with 2 dry cells and 1 foil ribbon.

(1) It lights:	(2) No light:	(1) It lights:	(2) No light:

c. Experiment with 2 bulbs plus 1 dry cell and 1 foil ribbon.

(1) Two shine bright.	(2) Two shine dim.	(3) One shines bright.	(4) No light.

2   Where are the two main contact areas. . .

a. . . .on a dry cell?	**b.** . . .on a bulb?

Objective

To further explore by trial and error the different ways to light a bulb with a dry cell and ribbon.

Introduction

These activities provoke a lot of curiosity about electricity. For this reason, you should issue a stern warning to your entire class: never fool with a wall plug, socket, or other electrical outlet. A severe shock or burn may result. Any experiment that goes beyond worksheet instructions must *first* have the teacher's approval.

On the other hand, what-would-happen-if questions are at the heart of scientific inquiry and should be encouraged. Assure the more timid in your classroom that the electricity produced by a few dry cells is too slight to even tickle. Wires may short circuit and get warm, but shocks are impossible to come by. Encourage your electricity-shy students to set their fears aside, turn their curiosity loose, and enjoy!

Lesson Notes

1. This activity is a continuation of the first. By trial and error, your students will discover there are many different ways to light a bulb with a cell and ribbon.

At times, students may be convinced a bulb should light when it doesn't. They'll likely complain that it is "burned out" or that the cell is "dead". This is usually not the problem.

The most common difficulty is a short circuit. The ribbon inadvertently touches a contact area where it shouldn't, causing the electrons to by-pass the higher resistance bulb filament completely.

Check Point

All contact points must be clearly drawn. Refer students, if necessary, to the schematic models provided in the last activity, step 6.

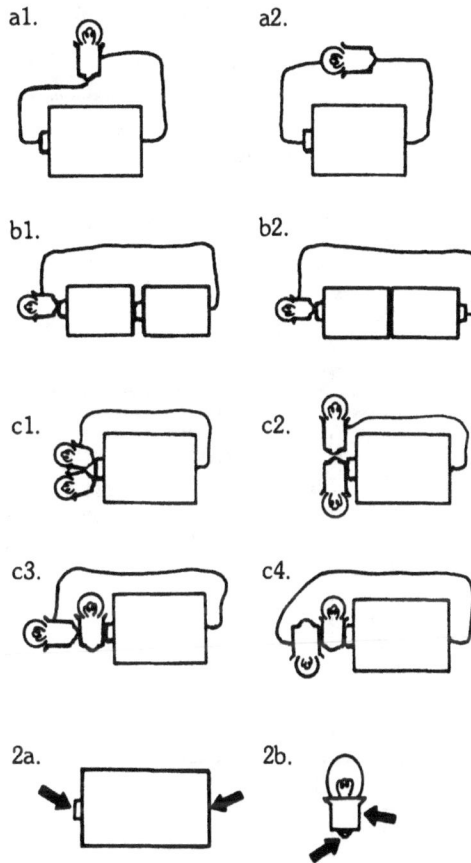

a1. a2.

b1. b2.

c1. c2.

c3. c4.

2a. 2b.

LIGHT BULB PREDICTIONS

1 In the tables below, first guess whether the dry cell lights the bulb.

Write your prediction next to each hook-up.

After you predict, experiment to see if you are right. Write each result in the tables.

REMEMBER HOW <u>MANY</u> CONTACT POINTS MUST CONNECT TO MAKE THE BULB LIGHT!

HOOK-UP	PREDICTION Will it light?	RESULT Did it light?	HOOK-UP	PREDICTION Will it light?	RESULT Did it light?
	a1.	a2.		d1.	d2.
	b1.	b2.		e1.	e2.
	c1.	c2.		f1.	f2.

IF YOU CAN'T HOLD ALL THE WIRES DOWN, ASK A FRIEND TO HELP!

2 You are now an expert on how to light a bulb. Write directions for someone who doesn't know how:

PSSST... REMEMBER TO TALK ABOUT CONTACT POINTS.

Objective

To use the idea of contact points to predict if a bulb will light. To test these predictions by experiment.

Introduction

None required.

Lesson Notes

The term "dry cell" (cell for short) is used in place of the word "battery" throughout these activities. A battery is defined as a collection of two or more interconnected cells. It is technically incorrect to refer to a single cell as a battery.

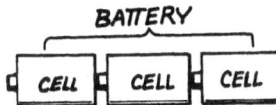

Common English usage, however, ignores this distinction. Even dry cell manufacturers refer to their products as batteries . Your students will too. That's okay. In time, as they continue to see the terms "dry cell" or "cell" in print on these worksheets, they'll make these words a part of their active vocabularies as well.

1. The order of doing things in this step is important. Students must (1) write down their prediction first, then (2) do the experiment, and finally (3) write down the result. Those who rush ahead and experiment *before* they predict miss the whole point: to apply the principle of contact points in a rational, predictive way to bulbs and cells.

In b and f, the bulbs do not light even though electricity flows through the ribbon. Electrons simply pass through the collar of the bulb (but not the filament) on their way from the negative to the positive pole. Because electrons are flowing and the cell is being drained of energy, your students should make these connections only briefly.

2. Be fussy about this answer. It's time for your students to summarize in their own words what they've learned in these last 3 activities, and in so doing, solidify concepts in their own minds. Encourage them to illustrate answers with diagrams.

Check Point

1a. yes/correct
b. no/correct
c. no/correct
d. yes/correct
e. yes/correct
f. no/correct

2. Contact points a or b on the bulb must touch, or be linked by wire, to poles x or y on the cell. The remaining two contact points (one on the bulb and one on the cell) must then be interconnected to complete the circuit and light the bulb.

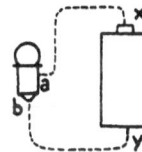

SERIES MEANS IN A ROW

1 Wrap a 30 cm foil ribbon around the collar of your bulb. Pinch the foil against the collar, then twist to hold the bulb tight. **TWIST**

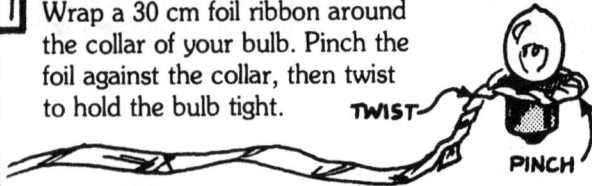

PINCH

2 Tape a paper clip to the side of your dry cell so it won't roll.

IT MAKES A *FOOT!*

3 If the bulb shines "dim" with 1 cell, find out how it shines with more cells connected in a *series*. Tell if it shines bright, medium or dim.

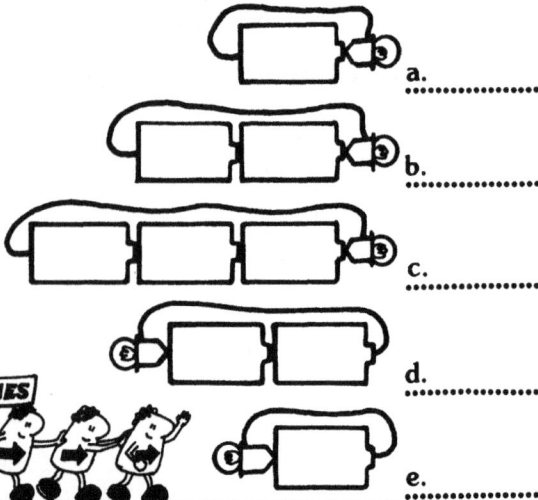

a. ....

b. ....

c. ....

SERIES

d. ....

e. ....

f. What happens as you add more cells in series?

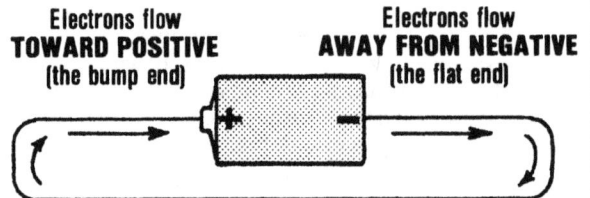

| Electrons flow **TOWARD POSITIVE** (the bump end) | Electrons flow **AWAY FROM NEGATIVE** (the flat end) |

g. Which way do electrons flow in circuits "a" through "e"?

CLOCKWISE   COUNTERCLOCKWISE

4 If the bulb shines "dim" with 1 cell, tell how it shines with more cells connected in *opposition* and *series*: bright, medium, dim or not at all.

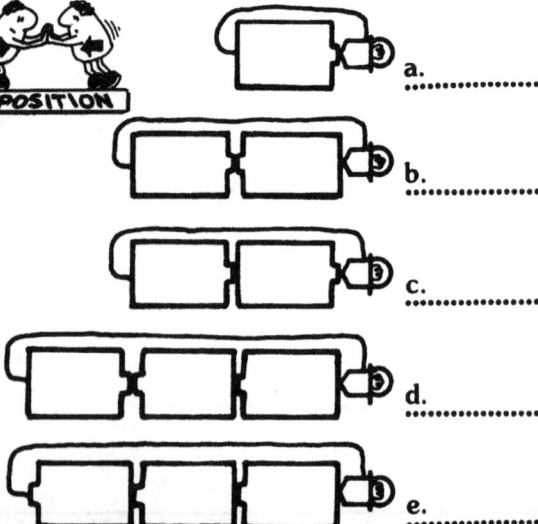

OPPOSITION

a. ....

b. ....

c. ....

d. ....

e. ....

f. Cells "**b**" are in opposition. What other cells are in opposition?

Cells "**c**" are in series. What other cells are in series?

g. Tell how electrons flow through each circuit; clockwise or counter clockwise.

a.   d.

b.   e.

c.

h. Why do cells "**d**" give less light than cells "**e**"?

Objective

To learn how to connect cells in series and opposition. To understand how this affects bulb brightness.

Introduction

Students need access to at least 3 dry cells in this activity. If extra cells are not available, organize lab groups by 3's so students can share equipment.

Lesson Notes

1. Students should use the same foil ribbon they made in activity H-1. The ribbon's length need only approximate 30 cm, about the length of this worksheet.

To secure a good electrical connection, make sure the foil is pinched above and below the collar of the bulb, then twisted firmly. Joined in this manner, the bulb and ribbon will be used together as one assembly from now on.

2. The idea here is to keep the dry cell from rolling off the table. When a paper clip is taped perpendicular to the length of the cell, it stabilizes the cell with a kind of "foot." Some students may tape the paper clip parallel to the length of the cell. This does not prevent it from rolling.

3. One cell is arbitrarily said to shine "dim". Given this reference, students should use relative terms like "medium" and "bright" to express how the brighter bulb intensities compare.

Physicists arbitrarily think of electric current as flowing from positive to negative. Hence, current is defined as the flow of positive charge. This definition is not very useful when applied to the world of cells, bulbs, and ribbons. The word "current" is therefore avoided throughout these activities, and replaced with terms like "electricity" or "flow of electrons." Because electrons are negative, they are shown as being *repelled* from the negative pole and *attracted* to the positive pole.

Your students may be interested to know how electrons flow inside the bulb. Tiny wires connect the side and end contact points on the exterior of the bulb to the light-producing filament inside. The electrons, of course, can flow in either direction through the bulb, depending on how the cell is connected.

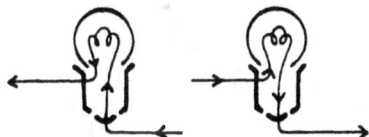

4. Your students may have difficulty determining the direction of electron flow in part (d). The 2 cells in series on the right have more push (voltage) than the single cell on the left, in opposition to them both. Electron flow is thus counterclockwise against the cell on the left. To a small extent, this cell is being recharged. See the "Extension" below.

These cells in part (d) actually give a little *less* light than the cell in part (a) due to the extra internal resistance of 3 cells in the circuit, instead of just one. A careful observer might write "extra dim" for part (d).

Extension

Electrons supplied by chemical reactions within the cell go around the circuit just once. A cell is dead when the chemicals that drive these reactions are about all used up. The cell can be recharged by using energy to drive the chemical reaction backwards, thus using up products to produce new reactants.

Suppose you want to recharge your car battery. Would you connect the charger in series (positive to negative) or opposition (positive to positive)?

Answer: Connect the charger to the battery in opposition; negative to negative and positive to positive. In this way, the charger can pump the electrons from positive back to negative.

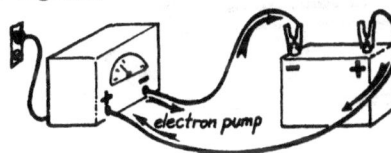

Check Point

3a. dim
 b. medium
 c. bright
 d. medium
 e. dim
 f. As you add more cells in series, the bulb shines brighter.
 g. All electrons flow clockwise.

4a. dim
 b. no light
 c. medium
 d. dim
 e. bright
 f. The left cell in part (d) is in opposition to the two on the right.
 The cells in (e) are all in series.
 g. (a) clockwise (b) no flow (c) clockwise (d) counter clockwise (e) counter clockwise
 h. The cells in (d) give less light because one is connected in opposition to the other two. All three cells in (e) are connected in series.

PARALLEL MEANS SIDE BY SIDE

1 Start with about 30 cm of foil ribbon attached to your light bulb.

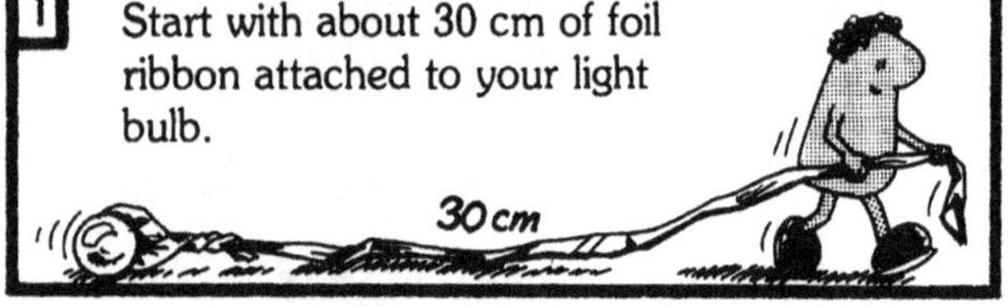

30 cm

2 Make a second foil ribbon about 20 cm long (as wide as this paper.)

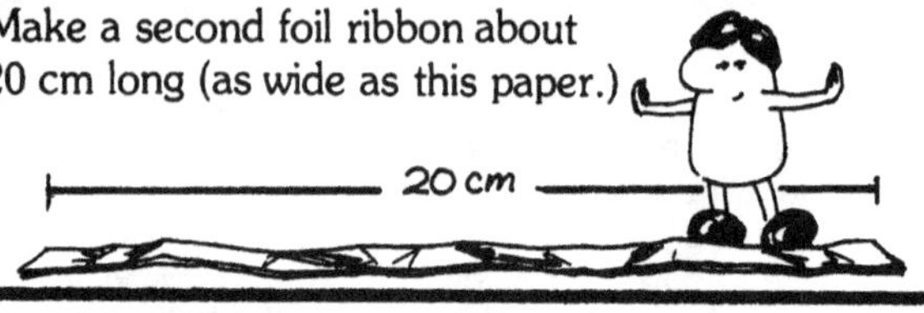

20 cm

3 If the bulb shines *medium* with 2 cells in *series*, find out how it shines with cells connected in *parallel*: bright, medium or dim.

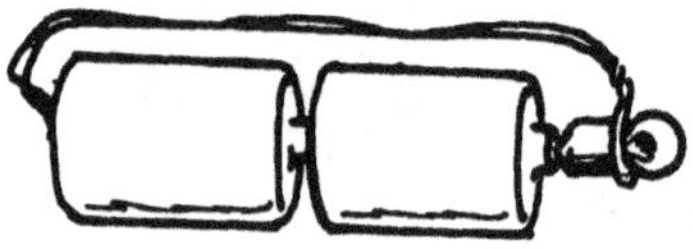

a.................

PARALLEL

b.................

SECOND RIBBON

c.................

d.................

e. Finish each sentence.

When you add more cells in parallel, the bulb. . .

To make a bulb shine brightest, it is best to connect. . .

4 Predict how each bulb shines: bright, medium, dim or not at all. Then experiment to see if you are right.

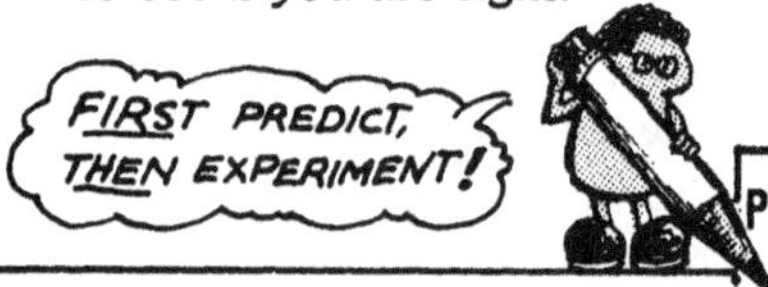

FIRST PREDICT, THEN EXPERIMENT!

	PREDICTION	RESULT
a. 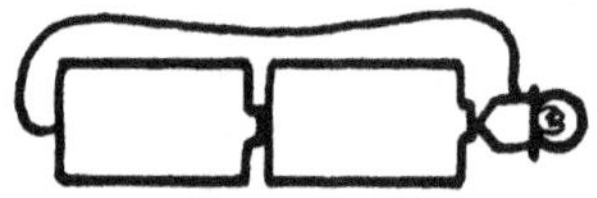	a1.	a2.
b. 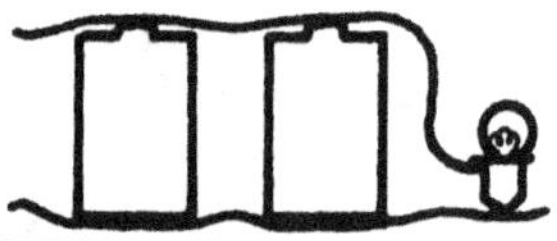	b1.	b2.
c. 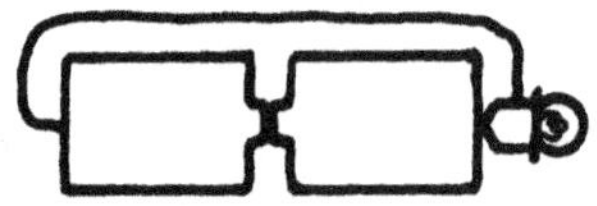	c1.	c2.
d. 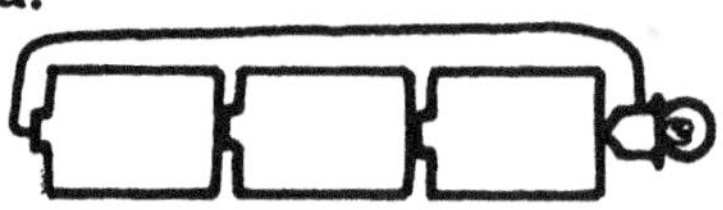	d1.	d2.
e. 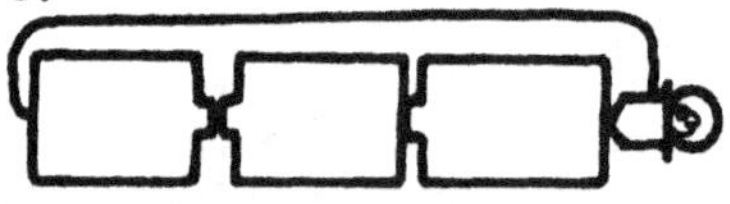	e1.	e2.

f. Tell how the cells are connected in each circuit.

a. *series*   d.

b.   e.

c.

Objective

To learn how to connect cells in parallel. To understand how this affects bulb brightness.

Introduction

Students need access to at least 3 dry cells in this activity. If extra cells are not available, organize lab groups by 3's so students can share equipment.

Lesson Notes

1-2. Students working together in lab groups may already have enough foil ribbons. If so, proceed directly to step 3.

3. Cells added in *series* increase *voltage* (the electromotive force carried by each electron); cells added in *parallel* increase *amperage* (the amount of electron flow). Bulb brightness is much more sensitive to increases in voltage than amperage.

Unless your students have had previous experience with bulbs and cells, this result is quite unexpected. Many will automatically assume that more cells give brighter light and erroneously write "dim", "medium," "bright," as the cells are added in parallel.

To overcome this preconceived notion, two cells in series at the beginning of this step are arbitrarily said to shine "medium." Through simple observation your students must admit that this medium intensity is much brighter than even three cells in parallel. Dim, dim, dim, therefore, is the most appropriate response.

The best way to compare bulb brightness is to stand three cells on a foil ribbon, as shown. Then complete the circuit with another bulb and ribbon touching first 1 cell, then 2 cells, then all 3. Put a coin under the bulb to make a steady glow (a better electrical connection).

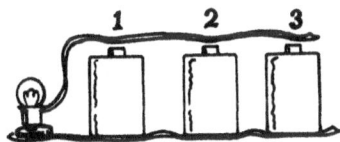

Students should observe that increases in bulb brightness are barely detectable as cells are added in parallel.

4. Look out for students who experiment first and don't predict. They are not exercising their minds, relating past experience to the problem at hand.

Praise those students who let a wrong prediction stand, even after they have come to a correct experimental result. Learning from one's mistakes is at the heart of scientific inquiry. Not admitting to bad predictions (erasing them) is bad science.

The bulbs in (b) and (e) both shine "dim." There is, however, a slight variation between them. Bulb (e) shines *very* dim because of the internal resistance of three cells in the circuit. Bulb (b) shines brighter (but much less than medium) because the circuit has less resistance and more electricity (amperage) — supplied by two cells, instead of one.

Check Point

3a. medium
 b. dim
 c. dim
 d. dim
 e. When adding cells in parallel the bulb remains dim.
    To make a bulb shine brightest, connect the cells in series.

4a. medium/correct
 b. dim/correct
 c. no light/correct
 d. bright/correct
 e. dim/correct
 f. (a) series (b) parallel (c) opposition (d) series (e) 2 cells in series in opposition with a third

CONDUCTOR OR INSULATOR?

1 Wind 3 rubber bands around your dry cell as *tightly* as you can.

FIRST
wind one
the long way.

THEN
wind two
the short way.

2 Wrap the free end of your ribbon and bulb 2 times around a coin. Slide this coin between the *flat* end of the dry cell and the rubber band.

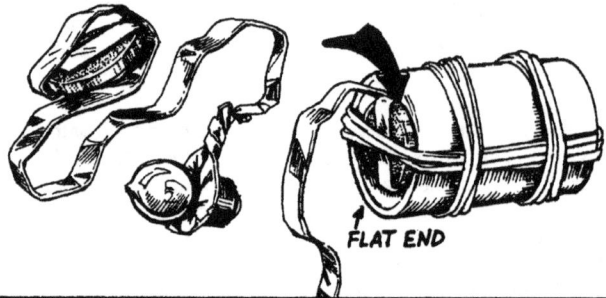

FLAT END

3 Slide the rubber band off the bump on your dry cell. Check to see that the bulb lights.

4 Use your bulb and dry cell to make a list of conductors and insulators. Fill in the table below.

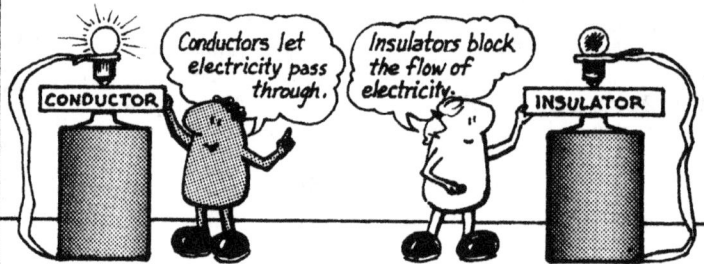

CONDUCTOR

Conductors let electricity pass through.

Insulators block the flow of electricity.

INSULATOR

a. WASHER? PENNY? GLASS? WOOD? CHALK

CONDUCTORS	INSULATORS
1.	1.
2.	2.
3.	3.
4.	4.
5.	5.
6.	6.
7.	7.
8.	8.
9.	9.
10.	10.

b. How are conductors alike?

c. Would foil ribbons work if we folded the tape to the outside and the foil inside? Explain.

Use CONDUCTOR and INSULATOR in your answer.

Objective

To use a bulb and dry cell to test if common materials in the classroom are conductors or insulators.

Preparation

Gather an assortment of objects in a box. Use things from your main inventory (foil, tape, thread, scissors, rubber bands, steel wool, nails) plus objects at hand (a washer, button, rubber stopper, tin can, candle, coin, piece of wood, etc.). About half of these items should be different metals — steel, copper, aluminum, brass, etc. The other half should be non-metals. Students will test materials in this box, plus other objects in your room to classify them as conductors or insulators.

Lesson Notes

1-2. Stretch the rubber band running lengthwise as tight as possible. It holds the coin in place. The other two rubber bands stretched around the circumference of the dry cell prevent this first rubber band from slipping off.

The coin guarantees an electrically secure connection. It presses the ribbon flat against the cell and serves as a conducting link between each terminal and the foil ribbon.

4. Some materials — the heating element on an electrical stove, or the filament in a light bulb, for example — fall between these two categories. These might be called poor conductors. Electricity passes through poor conductors, but with enough resistance to give off heat and light.

Conductivity is really a matter of degree. Given high voltages, skin is a conductor: an electrician takes every precaution to avoid touching a live wire. But with low voltages, skin is an insulator: Your students cannot pass electricity through their fingers to light a bulb — not even through a hangnail.

If you allow your students to wander about the room in this step, bulb and cell in hand, they will make wonderful discoveries. They may find, for example, that electricity will travel through a brass door knob or across a filing cabinet. (If it is painted, they must touch the bulb and cell to chipped places in the paint — usually along an edge — where the metal is exposed.) Given a long enough ribbon (students can twist them together), they may want to see if electricity will travel the length of your blackboard chalk tray. If it's made of wood, of course, they should be smart enough not to waste time trying. Those with silver teeth braces will become instant celebrities, being the only ones in the class with electric teeth.

Check Point

Answers will vary depending on the objects tested. Challenge older, more capable students to list *what* each object contains rather than the name of that object. The "conductors" column below, for example, lists only substances (more difficult). The "insulators" column lists only the name of each object (easier).

CONDUCTORS
aluminum
copper
nickle
brass
iron
steel
lead
tin
gold
silver

INSULATORS
eraser
toothpick
comb
window
candle
blackboard
chalk
skin
finger nail
rubber band

ELECTRIC PUZZLE

1   Fold 2 sheets of paper into quarters.

2   Draw 6 numbered circles on one of the papers like this.

3   Cut out each circle: bend the paper and cut into the folded edge.

4   Tape foil over the holes so some are connected and others are not.

*Use the **un**numbered side.*

KEEP SHINY SIDE UP

CONNECTED   UNCONNECTED

5   Cover with the other folded paper so no one can see which holes are connected.

6   Fasten with 2 paper clips. Write your name on the puzzle.

7   Unbend a paper clip. Slide it between the bump end of your dry cell and the rubber band.

8   Trade your puzzle for a friend's. Use your bulb and dry cell to find out which holes are connected with foil.

9   Record your results in a table.

WHOSE PUZZLE?	PREDICT WHICH HOLES JOINED.	RIGHT OR WRONG?
a.		
b.		
c.		

Remove the cover to see if you are right. . .

...then put the puzzle back together so someone else can try.

Objective

To construct a circuit puzzle. To find by trial and error those holes that are connected and those that are not.

Introduction

The purpose of this activity will not be completely clear to your students until they reach step 8. It is helpful, therefore, to make a circuit puzzle yourself. Show how different holes are connected by foil in your puzzle, then demonstrate how these connections are hidden between paper.

Don't tell your students how to solve this puzzle with the bulb and dry cell. (They should figure this out later on their own.) Deciding which holes are interconnected by foil is, of course, simply a matter of finding all combinations that light the bulb. Your students must exercise care to conduct a methodical search: there are 15 different double combinations to consider.

Lesson Notes

4. If different sides of the foil are placed face down, some holes will appear shiny and others dull. Holes that look identical are more likely (though not necessarily) interconnected. If only *one* side of the foil shows through the holes, as shown, this contextual clue will be eliminated.

7. A coin wrapped in foil ribbon connects the side of the cell you can't see. See steps 1 and 2 in activity H-6. 8. If your students are working through these activities at their own pace, those who reach this step first may not find anyone to trade puzzles with. You might be ready with a few puzzles of your own to trade. Here are some suggestions.

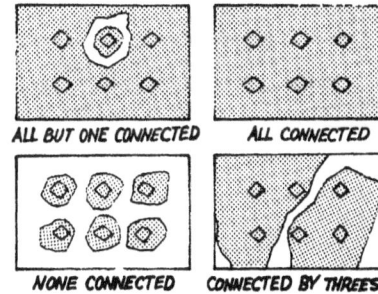

ALL BUT ONE CONNECTED   ALL CONNECTED

NONE CONNECTED   CONNECTED BY THREES

9. There is room in the table for each student to try solving the puzzles of at least three friends. If your students want to solve additional puzzles, by all means encourage them to extend this table.

Tell students to keep the foil ribbons attached to their bulbs; the rubber bands wound about their dry cells. They will use this equipment again, as is, in the next series of activities.

Check Point

9. Inspect each student's table to see if at least 3 puzzles have been attempted. Were some solved successfully?

I. CIRCUITS

ELECTRO-SQUARES MAP:

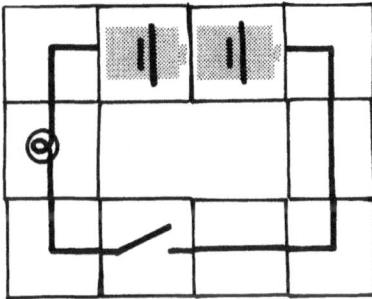

ACTUAL CIRCUIT:

As circuits increase in complexity, the number of loose ends that need holding soon exceed the fingers and thumbs available. Battery holders, bulb holders and switches to the rescue! Students improvise these circuit components from clothespins, coins, rubber bands and paper clips.

Students learn to plan their circuits first, before building them. Electro-squares help facilitate the planning process. These cut-out squares of paper already have standard circuit notation printed on them. Students arrange and rearrange these squares until they are satisfied with their overall circuit designs.

Soon your students will learn to diagram circuits directly on paper without the aid of electro-squares. They'll analyze series and parallel arrangements, deciding which applications are best for wiring buildings, just like electrical engineers.

EVALUATION

Each question evaluates a single activity from CIRCUITS as numbered. Use any combination to frame a formal exam or an informal review: Copy these questions on your blackboard, construct your own ditto master, or photocopy the questions while masking out the rest of the page. Evaluate in ways that suit your own teaching style, enabling your students to learn and enjoy science.

Questions

I-1
Describe how electrons flow through this circuit. Where do they start? Through what do they pass? Where do they end up?

I-2
This circuit has a by-pass at (a), another by-pass at (b), and an open switch at (c). Which of the following sentences are true?

1. The bulb lights if you close (c).
2. The bulb lights if you remove (a).
3. The bulb lights if you remove (b).
4. The bulb lights if you remove (a) and (b), then close (c).

I-3
Redraw this circuit using the correct symbols. Use arrows to show how electrons flow through the circuit.

I-4
Diagram this circuit:

1 cell + 1 switch
1 cell + 1 switch   } in parallel
2 bulbs in series

Use arrows to show how electrons flow through the wires. (It's OK to use electro-squares to help you solve this problem.

I-5 Compare each pair of circuits. Identify the one in each pair that produces more light. Circle one of the a's, one of the b's and one of the c's.

a1   b1   c1
a2   b2   c2

In which pair of circuits is the difference in the amount of light the greatest? Why?

I-6
a. Add 3 bulbs and 3 switches to this circuit so you can turn off each light independently.

b. Add 4 bulbs to this circuit so that if you unhook any one bulb, another will go off but the remaining two will stay on.

I-7
Which kind of circuit is best to put at the top and bottom of a stairway, a or b? Explain.

Answers

I-1
Electrons flow from the flat end (negative terminal) of the dry cell; through the ribbon to the collar of the bulb; out the bottom of the bulb, then through the coin to another foil ribbon; through this ribbon back to the bump end (positive terminal) of the dry cell.

I-2
Sentences (2) and (4) are true.

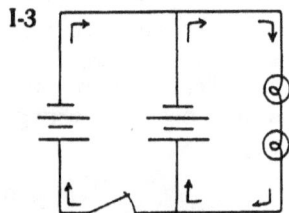

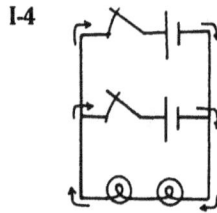

I-3

I-4

I-5
a1   b1   c1
a2   b2   c2

The difference is greatest in pair c. In c1, the bulbs are connected in parallel and the cells in series. Both factors combine for maximum brightness. In c2, the bulbs are connected in series and the cells in parallel. Both factors combine for maximum dimness.

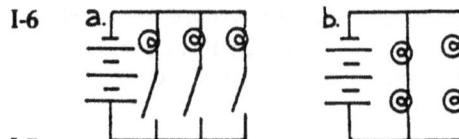

I-6 a.   b.

I-7
Circuit a is better because you can operate either switch independently. In circuit b, one switch can only function if the other switch remains in an on position.

SEQUENCING

CIRCUITS continues the study of electricity started in H. Students develop circuit components that are needed to complete J.

Related Activities: H—I—J

MATERIALS

Here is everything your students will use for the next 7 activities on CIRCUITS. Materials printed in normal type are part of the core 15-things-in-a-box inventory that support all 100 activities. Materials printed in *italics* are additional local materials that you provide or ask your students to bring from home. Pencil and paper are already assumed and therefore unlisted. Each item is numbered with the activity where it is first used.

(I-1) Clothespins.
(I-1) Flashlight bulbs. Use a size that is designed for 2 dry cells (3 volts).
(I-1) Foil ribbon from the "H" series of activities.
(I-1) Paper clips.
(I-1) Rubber bands.
(I-1) Masking tape.
(I-1) Aluminum foil.
(I-1) Scissors.
(I-1) *Small coins.*
(I-1) Size-D dry cells (1.5 volts).

FURTHER STUDY

Use problems like these plus "extension" ideas in CIRCUITS to lead your students beyond worksheet activity into original research and investigation. Each discovery leads to more questions, deeper questions, better questions than these. Answering them is what good science is all about.

Write an essay about life without electricity. What inventions and conveniences would we have to do without? How would this affect our quality of life?

Examine a piece of wiring used by electricians. How many individual wires does it contain? What does each wire do?

Design a floor plan for a house. Draw how you would wire each room.

NAME: _____   CLASS: _____

BUILD A CIRCUIT

1 Get 2 clothespin halves.

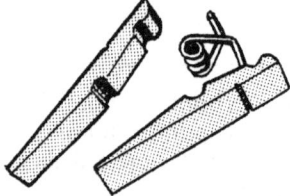

Clamp these around your bulb and foil ribbon.

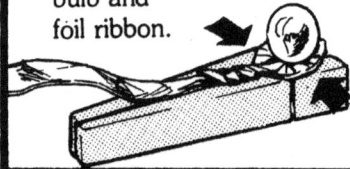

2 Place a paper clip so it touches the bulb and overlaps the collar.

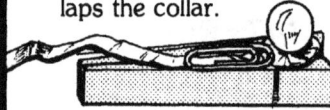

Secure with a rubber band.

Wrap TIGHT and EVEN.

3 Put a third clothespin-half under the first two. Wrap another rubber band around the end.

KEEP RIBBON FREE

Third half, flat side up.

Thick end under bulb.

4 Slip a coin between the bulb and clothespin. Slide a 20 cm foil ribbon under this coin.

5 Make 2 foil ribbons that are each 12 cm long.

12 cm is as long as 2 dry cells.

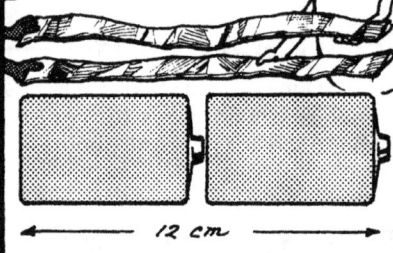

← 12 cm →

6 Wrap each ribbon 2 times around a coin, and slide it under the rubber band at each end of your dry cell.

7 Hook your bulb holder and dry cell together with paper clips.

I made a CIRCUIT.

PUT YOUR NAME ON YOUR BULB HOLDER

What path do electrons take as they move around your circuit?

Objective

To make a bulb holder and dry cell holder to use in later activities.

Introduction

Construct a bulb and dry cell holder yourself, before your students try. This will familiarize you with the directions, and provide a model for your students to follow.

Lesson Notes

1. These clothespin halves must be pulled apart just once. (You can throw the spring away.) Thereafter, you can recycle the halves to use again and again each time you teach this series of activities.

The foil ribbon has already been pressed around the collar of the bulb in previous activities. Make sure your students have pinched this foil to the collar of the bulb as tightly as possible (fingernails work well) so that no excess foil sticks up. This insures that the paper clip, bulb, and clothespin all fit snugly together in step 2.

2. The paper clip wrapped in the rubber band *must* overlap the collar of the bulb. This presses the bulb firmly against the top of the clothespin. It cannot slide when pushed from below by another clothespin half in step 3.

If your bulbs have screw-in threads instead of collars, omit this paper clip entirely. Instead wrap the rubber band directly around the clothespin as tightly as possible. If wrapped securely, the threads in each bulb's neck should bite into the wood of the clothespin, holding it firmly in place.

3. Wrap this 3rd clothespin half with the other two so the tapered ends of all three press together as illustrated. Wind a rubber band around these more loosely — just tight enough to hold the bottom clothespin half gently against the

bottom of the bulb. Keep the foil ribbon entirely free of this second rubber band.

4-5. Make the foil ribbons in the usual way. See activity H-1 steps 1 through 4.

4. This coin insures a good electrical connection between the bottom contact point of the bulb above and the foil ribbon below.

6. After the ribbon is wrapped two full turns around the coin, there will not be very much left to stick out beyond the edge of the cell. This is ideal. Ends that are too long tend to entangle and short out the cell. A good way to ensure long cell life is to clip long ribbons back on each end so they can't quite touch.

7. Remind your students to write their names on their bulb holders. They will use them many times in activities to come.

Check Point

Require students to prove to you that their bulb and dry cell holders are well made. The bulb should shine strong and steady when paper clipped to the dry cell. If students needs to press here or there with their fingers to keep their bulb from flickering, ask them to trace the poor electrical connection and fix it. Reliable circuit components are a prerequisite for success in activities that follow.

7. Electrons flow from the flat end of the cell (negative pole) through the ribbon, then through the penny, then through the bulb, and finally back through the other ribbon to the bump end of the cell (positive pole).

ELECTRIC BY-PASS

1 Make a switch: tape foil ribbons about 12 cm long (as long as 2 dry cells) to each end of a clothespin half.

OVERLAP THE ENDS

2 Bend the top ribbon so it touches the bottom ribbon *only* when you push down.

PUT YOUR NAME ON YOUR SWITCH

3 Build a circuit: connect your bulb, dry cell and switch with paper clips.

Tell how your switch is able to turn the light on and off.

4 By-pass your switch (in the *off* position) with another foil ribbon like this.

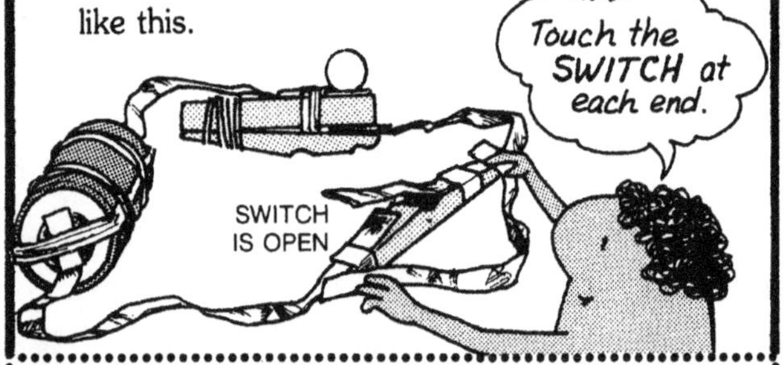

Touch the SWITCH at each end.

SWITCH IS OPEN

What happens to the bulb? Why?

5 Use another clothespin to keep your switch in the *on* position, then by pass the bulb.

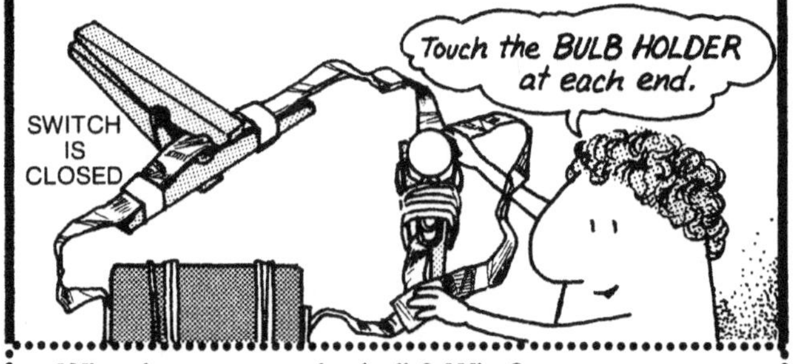

SWITCH IS CLOSED

Touch the BULB HOLDER at each end.

What happens to the bulb? Why?

Objective

To make a switch and integrate it into a simple circuit. To study how alternate pathways around the bulb and switch affect a simple circuit.

Introduction

None required.

Lesson Notes

1. If students separated 2 clothespins in the last activity, to make a bulb holder, then the remaining half clothespin can now be used to make the switch.

The foil ribbons must *not* be taped where they overlap. This will insulate the contact points, rendering the switch useless. Your students are less likely to make this mistake as long as you provide clearly visible masking tape, rather than hard-to-see clear tape.

2. Remind your students to write their names on their switches. They will use them again in later activities.

Extension

Here is a puzzle. Look under the hood of a car or truck. Notice that wires lead *out* from the positive terminal of the battery to all the lights in the vehicle. But no wires lead *back* from those same lights to the negative terminal. How can electricity flow in a complete circuit through such a system?

To solve this puzzle you and your students will need access to a vehicle with a 12 volt battery, plus good light (a flashlight, perhaps) to clearly illuminate battery cables and connections. In addition, you'll need a 12 volt car bulb plus about a meter of *insulated* wire that is thick enough to carry electricity from the battery to this bulb without overheating.

Peel back the insulation about 4 cm, far enough to wrap bare wire around the neck of the bulb. Expose about 1 cm of bare wire at the other end as well. The rest of the wire *must* be covered by insulation.

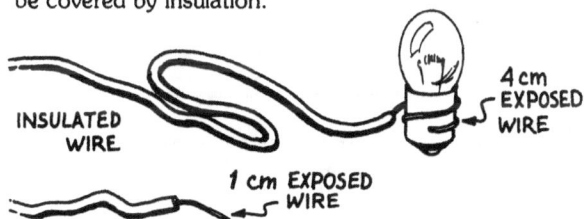

Demonstrate to your class how to light this bulb: touch the free end of the wire to the positive terminal of the car battery while you touch the bottom of the bulb to any part of the car frame that is not insulated from the engine chassis or covered by paint.

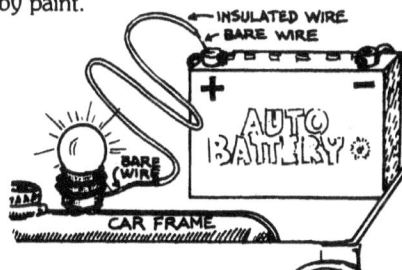

Ask your class to hypothesize how electrons can possibly travel from the negative terminal of the battery to the bulb. This mystery clears when you trace the cable that connects the negative terminal to the engine chassis: Electrons travel from the battery directly into the car frame! Touch the bulb to any place on the car frame and you've completed the circuit back to the positive terminal of the battery. Let there be light.

Caution: Never use bare wire. Touch any part of this wire to the car frame or engine by mistake, while an end is still connected to the battery, and sparks will fly in a spectacular short circuit. Use insulated wire only, touching only the insulation. A shocking experience may be in store for you if you provide a conducting path from the battery through your body to the ground!

Check Point

3. The bulb lights as electrons flow through it. When the switch is turned off, this flow is blocked. A gap is created over which the electrons cannot pass.

4. The bulb that was originally dark (because of an open switch) now lights up. The by-pass allows electricity to detour around the gap created by the open switch.

5. The bulb that was originally shining (because of a closed switch) now goes dark. Electricity flows through the by-pass ribbon instead of the bulb.

CIRCUIT SYMBOLS

1 Both drawings show the same circuit. Name each circuit symbol.

THIS ONE IS MUCH EASIER TO DRAW...

a. ....

b. ....

c. ....

2 Fill in each letter with the correct symbol.

A PUZZLE!

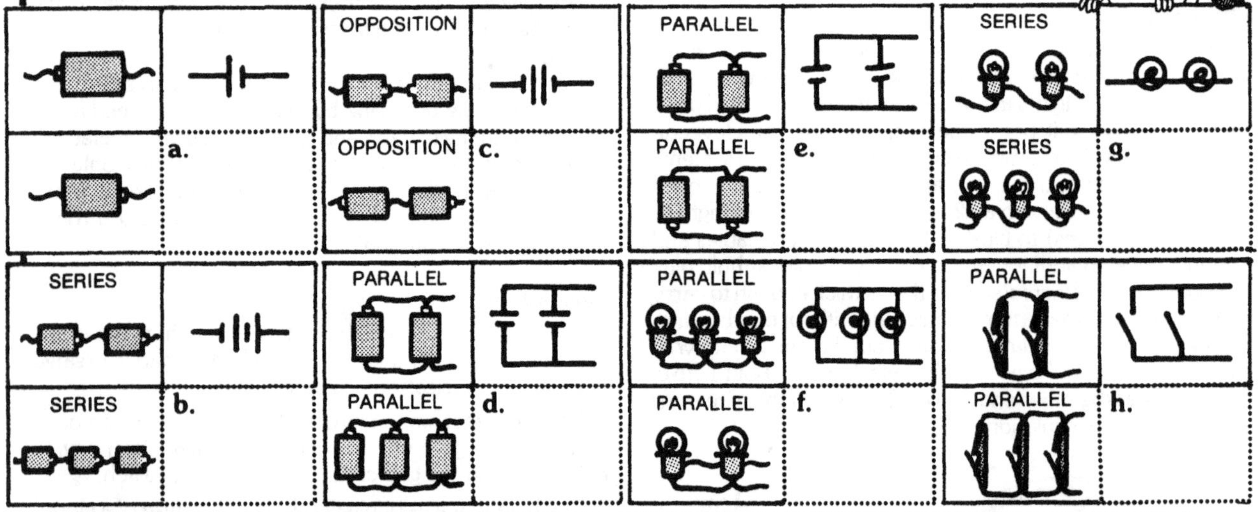

		OPPOSITION		PARALLEL		SERIES	
	a.	OPPOSITION	c.	PARALLEL	e.	SERIES	g.
SERIES	b.	PARALLEL	d.	PARALLEL	f.	PARALLEL	h.
SERIES		PARALLEL		PARALLEL		PARALLEL	

3 Redraw each circuit using the symbols you have just learned.

a.

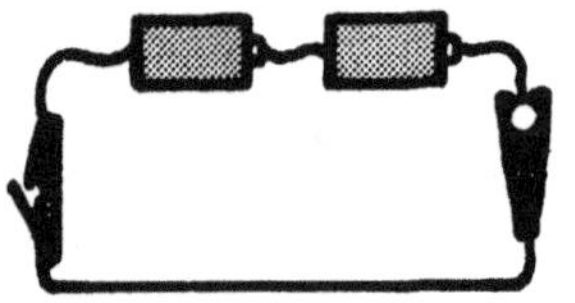

b.

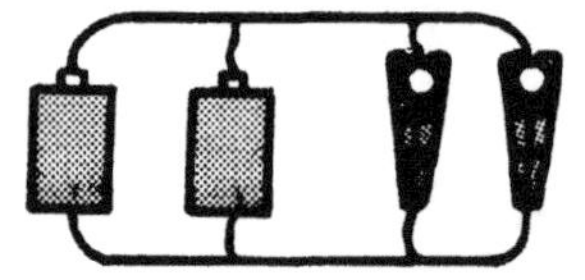

c.

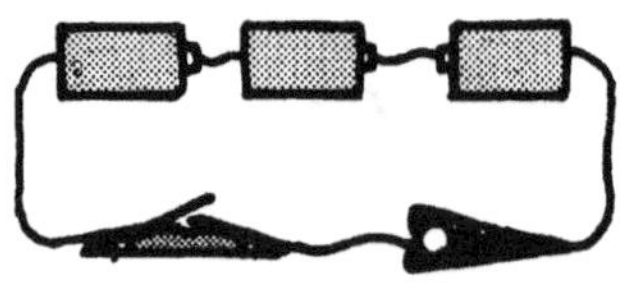

d. You can show a switch is closed by drawing an arc through it.

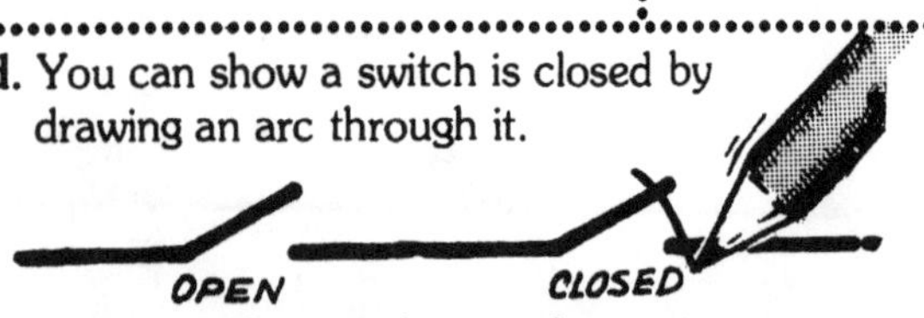

OPEN   CLOSED

"Close" the switch in each circuit you have just drawn.

e. Electrons always flow from negative to positive.

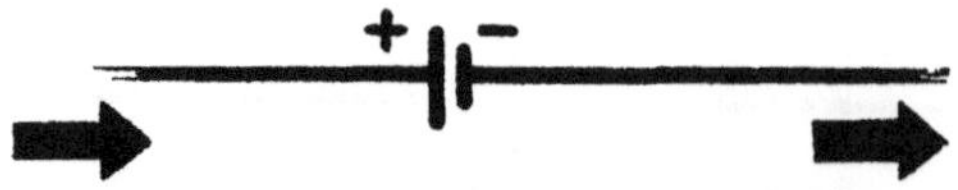

+ | | −

Use arrows to show how electrons move through each circuit you have just drawn.

Objective

To learn how to draw simple circuit diagrams using accepted symbols. To predict how electrons flow through these circuits.

Introduction

Scientists draw circuit diagrams to show how bulbs, switches and cells are interconnected. The symbols they use strongly suggest the form of the bulb, switch or cell they wish to represent. Your students will quickly master circuit diagramming, taking great delight in expressing the interconnectedness of things using their strange new language. Beyond pencil and paper, the only other requirement for this activity is a good eraser. As with any new language, students are bound to make errors that need correcting.

Lesson Notes

2-3. The symbol for cells seems most difficult to learn. Common errors include the following:

The long positive pole cannot be distinguished from the short negative pole.

Connecting wires between cells are drawn in.

The space between the poles is drawn too wide, as if the body of the cell should fit between.

A good way to remember that the longer line represents the positive pole is to recall the internal structure of a cell. In a zinc-carbon dry cell, the bump at the positive end caps a long carbon rod that extends down through the black electrolyte inside. Think of the longer line as representing this carbon rod. Since it *receives* electrons from the zinc can, it must be positive.

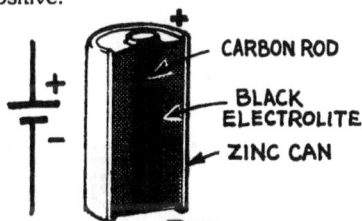

Check Point

1a. dry cell
 b. bulb
 c. switch

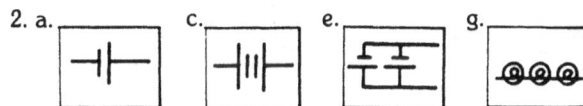

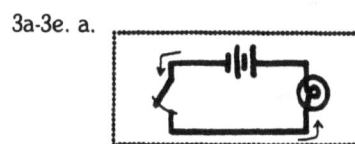

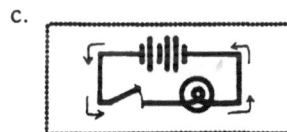

2. a.

3a-3e. a.

NAME: CLASS:

Circuits **I-4**

ELECTRO-SQUARES

1 Find the paper called
ELECTRO-SQUARES.
Cut out each square along
the dashed lines.

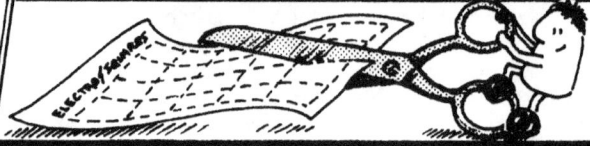

CUTOUTS
I-4

2 For each letter below:

Map the circuit with your
ELECTRO-SQUARES.

Draw the circuit.

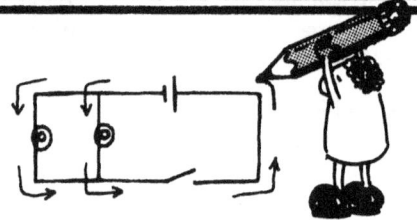

Show how electrons
flow through the circuit.

a. 1 cell + 1 bulb + 1 switch:
☐ Map
☐ Draw
☐ Show
    flow

b. 3 cells in *series* with
1 bulb + 1 switch:
☐ Map
☐ Draw
☐ Show
    flow

c. 3 bulbs in *parallel*
with 1 cell + 1 switch:
☐ Map
☐ Draw
☐ Show
    flow

d. 2 cells in *series*
+ 2 bulbs in *series*
+ 1 switch:
☐ Map
☐ Draw
☐ Show
    flow

e. 2 cells in *parallel*
+ 2 bulbs in *parallel:*
☐ Map
☐ Draw
☐ Show
    flow

f. 1 bulb + 1 switch } in *parallel* with
1 bulb + 1 switch } 2 cells in *series:*

☐ Map
☐ Draw
☐ Show flow

g. 1 cell
1 bulb } in parallel:
1 switch
☐ Map
☐ Draw
☐ Show
    flow

h. 2 cells in *opposition*
1 cell + 1 bulb } in *parallel:*
1 switch + 1 bulb
1 switch + 1 bulb
☐ Map
☐ Draw
☐ Show
    flow

3 Paper clip your
ELECTRO-SQUARES
into one pile. You'll
need to use them
again.

Objective

To practice mapping and drawing more complicated circuit diagrams. To predict how electrons flow through these circuits.

Introduction

None required.

Lesson Notes

1. Sheets of electrosquares are available as student cutouts. You can also reproduce them, if you wish, from the line master in the back of this book.

2. Three Teacher-Check squares in each box emphasize 3 distinct tasks (map/draw/show flow) to be completed for each circuit. If you check off these squares for your students, they will probably work harder and faster. Checks are motivational, much like a word of praise or pat on the back. More independent students, of course, can easily monitor their own progress with self-checks.

Students should again use pencils so they can erase errors and draw neat diagrams. Encourage students to draw rectangles with sharp corners; to indicate electron flow with just a *few* arrows drawn beside the wires, not on the wires.

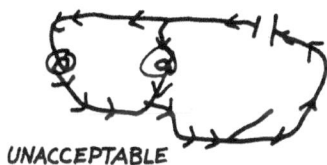

UNACCEPTABLE

To familiarize students with the dry cell symbol printed on some electro-squares, the actual shape of the cell is shaded into the background. When students diagram each circuit after mapping it, they should copy only the symbol for the cell, not its background image.

These electro-squares serve as an intellectual crutch, helping students to piece together circuits they otherwise would not be able to draw. As they progress through this lesson, some may set this crutch aside and begin to walk unaided. This is ideal. Allow students to skip the mapping part of each problem if they are able to diagram correct circuits directly.

Notice that students don't actually build these circuits with bulbs and dry cells. Not yet. Circuit building begins in the next activity, *after* students have first learned to diagram them.

2a. To avoid excessive wordiness in these problems, terms like "1 cell + 1 bulb + 1 switch " are assumed to mean "in series," unless otherwise stated.

2b. Here, the electro-squares need to be used wisely. If students try to line up the cells, bulb, and switch all on the same side, they will run out of squares.

2c. The three bulbs need to be connected in parallel, not series. Students who experience difficulty should consult the picture directly above this problem, showing 2 bulbs in parallel. As an added hint, you might suggest that students concentrate on using "T" shaped electro-squares when attempting parallel constructions.

Check Point

Cells, bulbs and switches may be sequenced differently from these and still contain the right number of circuit components organized in the correct series or parallel configuration.

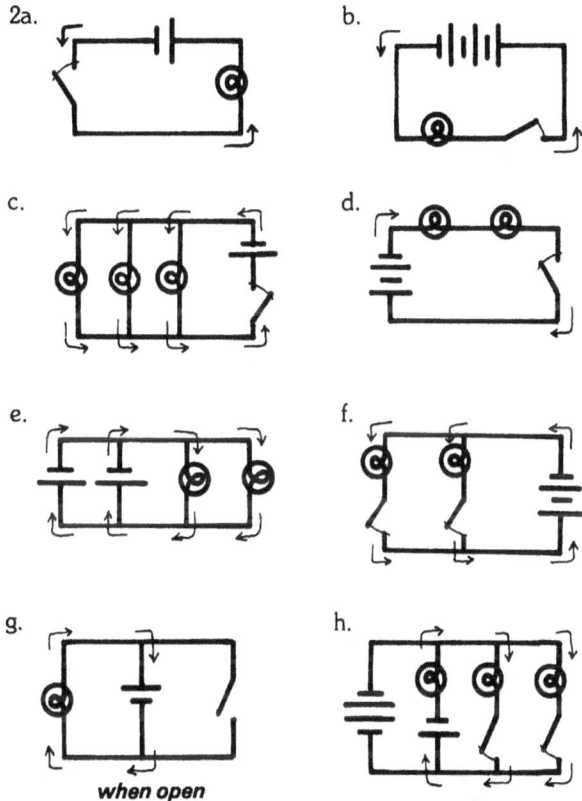

2g. In this circuit, electrons flow through the bulb only when the switch is open. Close the switch and the bulb is bypassed. Your students will actually build this "backwards" circuit in activity J-3.

2h. There must always be a balance between the number of electrons leaving and entering a cell. Electrons cannot flow *out* from both cells in opposition, for example, without also flowing back *in*. These opposition cells, therefore, must remain inactive.

MAP IT/DRAW IT/BUILD IT

Map each circuit below with your ELECTRO-SQUARES.

Draw each circuit.

Build each circuit. *Compare* how bright the bulbs shine.

1a. *2 CELLS in series* with 1 bulb + 1 switch in series.
☐ Map it
☐ Draw it
☐ Build it

1b. *2 CELLS in parallel* with 1 bulb + 1 switch in series.
☐ Map it
☐ Draw it
☐ Build it

1c. Which way of connecting *CELLS* makes the bulb shine more brightly?

2a. *2 BULBS in series* with 2 cells + 1 switch in series.
☐ Map it
☐ Draw it
☐ Build it

2b. *2 BULBS in parallel* with 2 cells + 1 switch in series.
☐ Map it
☐ Draw it
☐ Build it

2c. Which way of connecting *BULBS* make them shine more brightly?

Objective

To learn how to connect dry cells to achieve maximum brightness.

Introduction

Your students have correctly diagrammed a circuit and build it according to plan (or so they think). But the bulbs still won't light. Do they run to the teacher for help? Only as a last resort. Here is a simple check list to help students solve circuit problems independently. (Write the capitalized summary of these 3 steps on your blackboard. Explain the details in a class discussion.)

1. CHECK CELL DIRECTIONS

If the cells are in series, you must connect *unlike* poles together — positive to negative. Otherwise you have connected them in opposition.

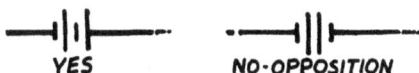

If the cells are in parallel, you must connect *like* poles — positive to positive and negative to negative. Otherwise you have connected them into an energy-draining short circuit. They won't last long connected like this.

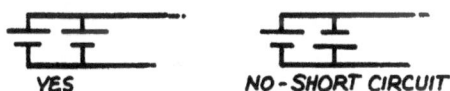

2. CHECK CONNECTIONS

Do paper clips press both foil ribbons firmly together at *every* junction?

3. CHECK COMPONENTS

Test each component in the circuit with other equipment that you *know* is working.

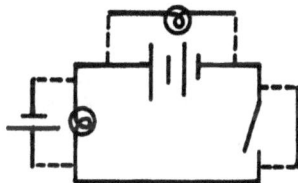

Your class should now be ready to work through this activity independently. Students need to share equipment to make at least 2 cells and 2 bulbs available per activity group. Leave these trouble-shooting steps on your blackboard for students to use as a reference.

Lesson Notes

Teacher-Check squares are again provided to insure that students complete all parts of one problem (map/draw/build) before proceeding to the next. These squares can also be self-checked by students themselves, unless you wish to monitor student progress more closely.

Depending on the ability of your students, you might elect to make circuit mapping an optional procedure. However, those who continue to draw circuits incorrectly should be encouraged to map their way to the correct answer.

1-2. When comparing series configurations in part (a) with parallel configurations in part (b), it is important to control variables. Use the *same* dry cells and bulbs throughout. Further, both bulbs in step 2 should have the same resistance and be manufactured by the same company. This insures that the bulbs will shine with equal intensity in each circuit.

Check Point

1a. 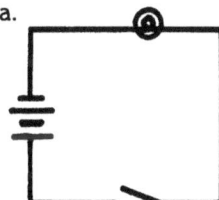   1b.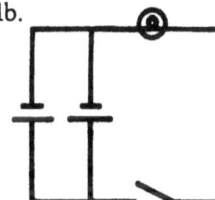

1c. Cells in series make the bulb shine brighter than cells in parallel.

Cells in series increase voltage — the electromotive force carried by each electron. Cells in parallel increase amperage — the amount of electron flow. This is analogous to water over a dam. To make a water wheel spin more rapidly, you would do better to raise the dam (increase the number of volts) rather than add more water (increase the number of amps).

2a. 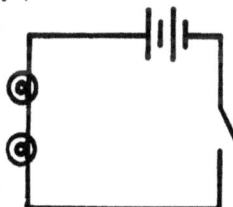   2b.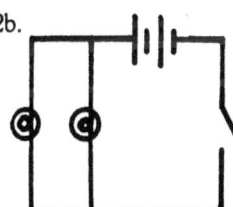

2c. Bulbs in parallel shine brighter than bulbs in series.

Connected in series, each bulb must share the voltage delivered by the cells. Connected in parallel, each bulb has an independent path to the cells and thus receives full voltage. Using the same analogy, water wheels spin faster in the falls when placed side-by-side (in parallel). If the wheels are positioned so that one is directly over the other (in series), they will both use the same water and thus spin more slowly.

SERIES OR PARALLEL?

1a. Draw and build this circuit.

1 bulb + 1 switch
1 bulb + 1 switch } in SERIES with 2 cells in series

1b. Do the bulbs shine brightly or dimly?

1c. Will one switch turn on one light? Why?

1d. If a bulb burns out, will the other still shine? Why?

2a. Draw and build this circuit.

1 bulb + 1 switch
1 bulb + 1 switch } in PARALLEL with 2 cells in series

2b. Do the bulbs shine brightly or dimly?

2c. Will one switch turn on one light? Why?

2d. If one bulb burns out, will the other still shine? Why?

3 Suppose you're an electrician. Would you wire a house in series or parallel? Give 2 reasons.

4 Suppose you make Christmas tree lights:

a. Is it easier and cheaper to wire them in series or parallel?

b. Which way of wiring makes the best product?

Objective

To understand why electricians wire buildings in parallel, rather than in series.

Introduction

This activity applies lessons learned about bulbs and dry cells to wiring in a house. To insure that students don't get the wrong idea, stress this important point once more: experimenting with household electricity is foolish and dangerous. An AC/DC 110 or 220 volt wall socket must never be equated with a dry cell.

Remind students, as well, that they are responsible for fixing circuits that don't work. The teacher should only be consulted as a last resort. If necessary, review the trouble-shooting checklist presented in the introduction to I-5. Keep these 3 steps written on your blackboard for students to use as a reference.

Lesson Notes

1-2. Each circuit demonstrates some of the advantages of parallel wiring over series wiring. Once assembled, students should leave these circuits intact until they have fully answered all the related questions. Those who pull them apart prematurely may miss important conceptual con-nections as well.

4. Your students may wonder how Christmas tree lights are wired together into what appears to be a single wire. An electric cord that is wired in parallel actually has two insulated wires enclosed in one single sheath.

CHRISTMAS TREE LIGHTS
IN **PARALLEL**

CHRISTMAS TREE LIGHTS
IN **SERIES**

Check Point

1a. This circuit is relatively easy to build. It hooks together in a large, continuous circle.

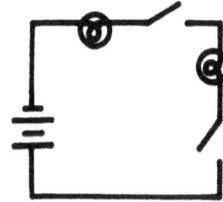

b. The bulbs shine dimly.

c. No. One switch in series can only turn on both lights, as long as the other switch remains closed.

d. No. The burned-out bulb would break the circuit for the other bulb as well.

2a. Because of its parallel construction, this circuit is harder to build. Students who experience difficulty here should first map it with their electro-squares.

b. The bulbs shine brightly

c. Yes. Each bulb has an independent path to the cells through its own switch.

d. Yes, The other bulb still has an unbroken, independent path to the cells.

3. Wire the house in parallel. Each appliance or light receives full power. Each can be turned on or off without affecting the others. If one burns out, the others all remain on.

4a. Series wiring is cheaper and easier.

b. Parallel wiring makes the best product.

NAME: CLASS:

TWO-WAY SWITCHES

1 Cut out 8 strips of foil about this big. ➔

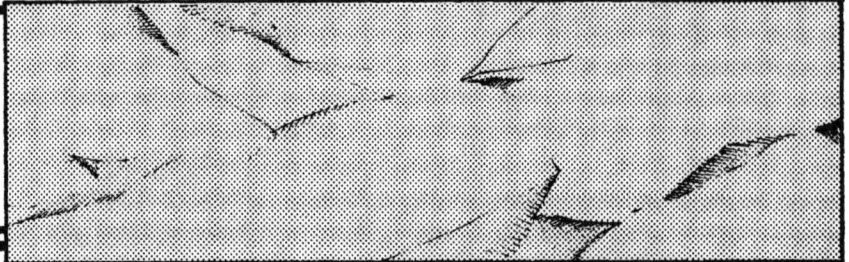

2 Wrap and tape the ends of 2 clothespins in the foil strips.

Most foil should stick out.

TAPE HERE

3 Wrap all 8 ends. *No* foil should touch the spring in the middle.

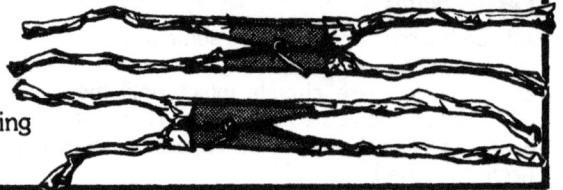

4 *Gently* twist the foil strips together like this. Be sure the clothespins point opposite ways.

5 Connect your 2-way switch in series with a bulb and 2 dry cells.

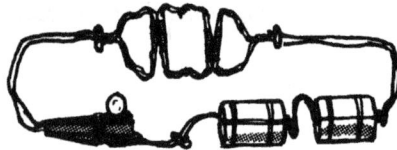

Why would 2-way switches like these be good to use at the top and bottom of a stairway?

6 This 2-way switch is turned on. . .

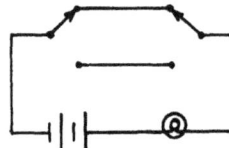

...Draw a *different* way to turn it on.

7 This 2-way switch is turned *off*. . .

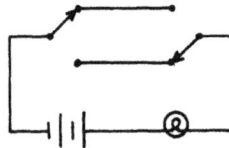

...Draw a *different* way to turn it off.

8 *****Have a contest.***** Ask a friend at one switch to keep the light turned off, while you try to keep it turned on at the other end.

Objective

To build 2-way switches and integrate them into a circuit. To understand how they work.

Introduction

A 2-way switch looks like this.

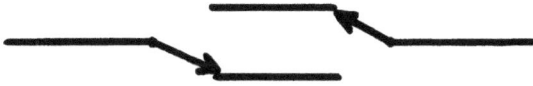

Flip the left side up, and the right side independently turns on (up) and off (down). Flip the left side down, and the right side still independently turns on (down) and off (up), this time in reverse.

A pair of simple switches cannot duplicate this special property. Connected in series, *both* switches must be turned on in order to compete the circuit.

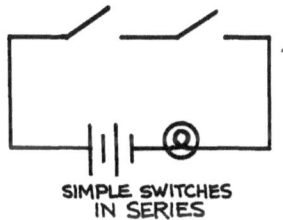

SIMPLE SWITCHES
IN SERIES

Connected in parallel, *both* switches must be turned off in order to break the circuit.

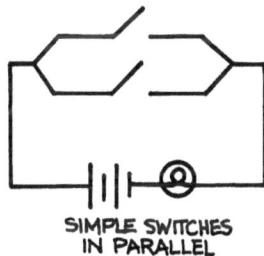

SIMPLE SWITCHES
IN PARALLEL

Lesson Notes

2-3. The foil must not touch the metal spring of either clothespin. Otherwise, a conducting path through the spring may develop, allowing electricity to pass along the length of the clothespin. If any 2-way switch remains on in either position, this has likely happened.

4. The watchword in this step is "*gently*". Your students are using foil that has not been strengthened with a tape backing. Twist too hard and the foil easily tears apart.

8. This little game of you-keep-it-on and I-keep-it-off is a favorite for children to play wherever 2-way switches are found — as long as there are no adults around to yell at them for wearing out the light switches. Here your students can play the game openly, wearing out their fingers before they ever break the clothespins.

The person competing to keep the switch turned off, however, must not cheat. If either clothespin is squeezed to a *midway* position, where the switch is neither turned on nor off, the circuit remains broken with no possibility of the other player reestablishing contact at the opposite end.

Check Point

5. The light can be turned on or off independently at both the top and bottom of the stairway. It doesn't matter whether the unused switch is turned up or down.

6.

7.

J. ELECTRICAL RESISTANCE

Common steel wool, the stuff you use to scrub pots and pans, is an ideal material for studying resistance to electricity. Thin steel fibers carry the electrical load of several dry cells, but their high resistance noticeably dims any bulb in the circuit.

When students place this same electrical load across fibers of steel wool that are especially short and thin, they will observe an even more dramatic demonstration of electrical resistance. The tiny steel strands literally flash, pop and melt — a miniature fireworks show.

Even fuses are possible. If your students select a steel wool fiber with the right diameter (some trial and error experimentation is necessary), it will carry the load and sustain a light bulb until a short circuit occurs. The resulting electrical mini-surge melts the fiber and shuts down the circuit, just like a real fuse.

——— EVALUATION ———

Each question evaluates a single activity from ELECTRICAL RESISTANCE as numbered. Use any combination to frame a formal exam or an informal review. Copy these questions on your blackboard, construct your own ditto master, or photocopy the questions while masking out the rest of the page. Evaluate in ways that suit your own teaching style, enabling your students to learn and enjoy science.

Questions

J-1
Six copper wires, 3 thin and 3 thick, have different lengths, as shown. Circle the wire with the greatest resistance. Draw a box around the wire with the least resistance.

THIN WIRES: *THICK WIRES:*

_____   _____

_____   _____

_____   _____

J-2
To save money, an electrician installs copper wire in a home that is thinner than the building code allows. Why is this a dangerous practice?

INSULATED WIRE Building Code Approved

BUDGET WEIGHT   *HIGHER COST*

J-3
Fill in the blanks with "on" or "off."

To make both lights shine, turn L _____ and R _____

To make one light shine, turn L _____ and R _____

To make no lights shine, turn L _____ and R _____

J-4
It's a hot summer evening. All the kitchen lights are on, and several fans are humming. You put your dinner in the oven and turn on the temperature control. Suddenly everything goes dark. What's wrong? What should you do?

Answers

J-1

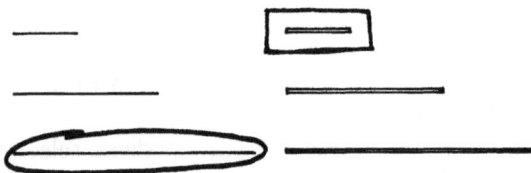

J-2
Thin wire has higher electrical resistance. It creates a serious safety hazard because the wires could easily overheat and cause a fire.

J-3
To make both lights shine, turn L *off* and R *off*.
To make one light shine, turn L *off* and R *on*.
To make no lights shine, turn L *on* and R *off or on*.

J-4
The lights and fans were already drawing maximum power through the line. Turning on the oven created an overload that blew a fuse. You should turn off the oven, get a flashlight, and go change the fuse. Then, before heating up your dinner, turn off the fans and some of the kitchen lights.

SEQUENCING

ELECTRICAL RESISTANCE completes the study of electricity started in H and continued in I. J should only be scheduled after H and I are completed.

Related Activities: H—I—J

MATERIALS

Here is everything your students will use for the next 4 activities on ELECTRICAL RESISTANCE. Materials printed in normal type are part of the core 15-things-in-a-box inventory that support all 100 activities. Materials printed in *italics* are additional local materials that you provide or ask your students to bring from home. Pencil and paper are already assumed and therefore unlisted. Each item is numbered with the activity where it is first used.

(J-1) Steel wool.
(J-1) Clear tape.
(J-1) *Scratch paper.*
(J-1) Paper clips.
(J-1) Dry cell holders, bulb holders and foil ribbon from the "I" series of experiments.
(J-2) Straight pins.
(J-3) Switches from the "I" series of experiments.
(J-4) Scissors.
(J-4) *Small coins.*
(J-4) Clothespins.

FURTHER STUDY

Use problems like these plus "extension" ideas in ELECTRICAL RESISTANCE to lead your students beyond worksheet activity into original research and investigation. Each discovery leads to more questions, deeper questions, better questions than these. Answering them is what good science is all about.

Divide your class into student teams. Debate the best way to generate electricity. Each team should champion a particular energy source like solar, geothermal, wind, synthetic fuels or nuclear. Have at least one group advocate conservation.

Read about electricity as it relates to other scientific disciplines.
● chemistry: reactions that make electricity (oxidation/reduction)
● biology: electric potentials within the body
● computers: integrated circuits, microchips and semiconductors
● physics: low temperature superconductors

Investigate one of these career possibilities in the field of electricity:

electrical engineer
electrician
electrical draftsperson
electronics technician

RESISTANCE IN A WIRE

1 Pull off a single strand of steel wool.

CHOOSE A *FAT* ONE.

2 Pull it straight and tape each end to a piece of paper.

Tape ends only.

3 Unbend a paper clip and tape it the same way.

4 Unbend another paper clip. Slide it under the coin on your bulb holder, replacing the foil ribbon.

5 Connect your bulb to 2 cells in series. Attach a foil ribbon and paper clip to the other end.

6 Pass electrons through the *thick* paper clip and the *thin* steel wool. Compare how bright the bulb shines.

THICK WIRE   **THIN WIRE**

FILL IN EACH SPACE WITH THESE WORDS

7 Pass electrons through the steel wool for a *long* distance and a *short* distance. Compare how bright the bulb shines.

LONG WIRE   **SHORT WIRE**

a. Shines brighter: ....

b. Shines dimmer: ....

c. Holds back flow of electrons: ....

d. Allows electrons to flow: ....

e. Has low resistance: ....

f. Has high resistance: ....

a. Shines brighter: ....

b. Shines dimmer: ....

c. Holds back flow of electrons: ....

d. Allows electrons to flow: ....

e. Has low resistance: ....

f. Has high resistance: ....

8   **a.** Resistance in a wire *increases* as. . .

(name 2 ways)

b. Resistance in a wire *decreases* as. . .

(name 2 ways)

Objective

To understand how length and diameter affect electrical resistance in wire.

Preparation

Success in this activity depends on students selecting steel wool fibers with the correct diameter. (They come in many sizes.) Run through these steps in advance to see what fiber size from your particular brand of steel wool works best. Will a "fat" steel wool fiber chosen in step 1, dim the bulb in steps 6 and 7 without melting? If so, all is well. If not, keep experimenting with thinner or thicker fibers until you find a better size that works for your particular brand of steel wool. Save one as a standard of comparison to help your students select fibers of the right diameter when they do the experiment.

All 4 activities in this series require 2 dry cells to work properly (3, if the cells are weak). Organize students in lab groups of 2 or 3 so they can pool equipment.

Lesson Notes

1. Exercise care when plucking fibers from the ball of steel wool. Careless handling may result in an occasional painful steel sliver. Make sure your medical kit contains a pair of tweezers, just in case.

4. This straightened paper clip replaces the bulb holder's foil ribbon in both this activity and the next.

5. It is important that the bulb shine reasonably bright when these end paper clips are touched together. If it doesn't shine at all, the cells are likely connected in opposition. If it shines only dimly, either the connections are bad or the cells are weak. First check the connections, then if the bulb is still dim, add one more cell in series to make a total of three.

6. The steel wool may melt if the paper clip probes are placed too close together. If this happens start over with another strand that is a bit thicker, and keep the probes a little further apart.

6-8. The observations about bulb brightness (in steps 6 and 7) should lead your class (in step 8) to logical conclusions about electrical resistance. Students who still don't fully understand may require additional concrete experience. The extension that follows is ideal.

Extension

Electrical resistance in wire is analogous to forcing air through a tube. It is easier to blow air through a wider tube than through a narrower tube; through a shorter tube than a longer tube.

Cut a piece of paper in 2 halves. Roll one of these halves tightly around your pencil and secure with tape. Roll the other half into a wider tube and tape it too.

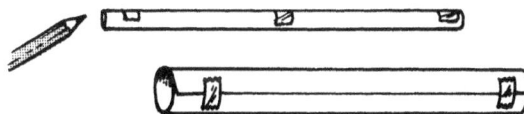

Cut a very short 2 cm piece from the end of the narrow roll.

Now, blow air through each tube as hard as you can. Which offers more resistance . . .

a. the *wide* long tube, or the *narrow* long tube? (narrow)
b. the narrow *long* tube or the narrow *short* tube? (short)

These results for pushing air through a tube are analogous to forcing electrons through a wire.

Check Point

6a. thick wire
 b. thin wire
 c. thin wire
 d. thick wire
 e. thick wire
 f. thin wire

7a. short wire
 b. long wire
 c. long wire
 d. short wire
 e. short wire
 f. long wire

8a. Resistance in a wire *increases* as thickness decreases and length increases.

8b. Resistance in a wire *decreases* as thickness increases and length decreases.

NAME: CLASS:

A FLASHY EXPERIMENT

1 Connect your bulb to dry cells in series. Attach a foil ribbon and straight pin to the other end.

2 Gently pull on a piece of steel wool so iron fibers fall on a piece of paper.

BE CAREFUL NOT TO GET SLIVERS IN YOUR FINGERS.

3 Pass electricity though the tiny steel wool wires on your paper. Observe what happens to the steel wool *and* the light bulb.

a. What happens when electricity passes through very *thin* wire?

b. What happens when electricity passes through *thick* wires or clumps?

c. Does *thin* wire resist the flow of electricity more than *thick* wire? Explain.

4 Order these materials by how much resistance they have:

BULB FILAMENT

THINNER STEEL WOOL STRAND

THICKER STEEL WOOL STRAND

FOIL RIBBON

1 ....
2 ....
3 ....
4 ....

HIGHEST

LOWEST

Objective

To watch a miniature fireworks show! To appreciate that high electrical resistances creates heat and light.

Introduction

Push your hands together hard. Rub them *slowly* back and forth. Feel the resistance, the friction created by sliding one hand past the other.

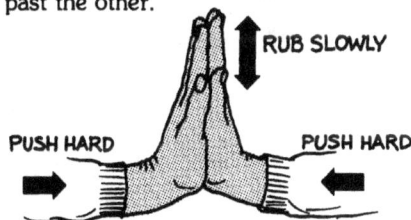

Push your hands together hard once more. This time rub them *rapidly* back and forth as fast as you can. The resistance (or friction) is much greater than before. It generates heat that you can feel. The greater the resistance, the more heat you can produce.

Electrons resist moving through a wire just like your hands resist moving past each other. This resistance causes the element on an electric stove to heat up and glow. This resistance lights the filament in an electric light bulb. Push enough electrons through a wire with enough force, and the resulting resistance will heat the wire, cause it to melt and even burn!

Lesson Notes

1. The bulb holder should have a straightened paper clip (instead of a foil ribbon) extending from beneath the penny. Students made this exchange in the previous activity.

For this experiment to be really flashy, the bulb should shine reasonably bright when the pin and paper clip probes are touched together. If fresh cells are used, 2 in series will work fine. If the cells are weak, 3 in series may be required.

2-3. Remind students once again to handle steel wool cautiously. Your medical kit should contain a pair of tweezers in case anyone needs to remove a sliver.

Rubbing steel wool back and forth on itself causes fibers of all sizes to settle on the paper. The very fine dust-like particles will literally disappear in a flash of light when electricity is applied. Mid-sized particles flash and melt, occasionally burning tiny holes in the paper and causing puffs of smoke. The heavier fibers actually carry electricity and light the bulb.

In general, the closer together you place the pin and paper clip probes, the more dramatic the display. This concentrates a maximum flow of electricity through minimum lengths of steel wool.

Check Point

3 a. The tiny fibers tend to flash and melt, without turning on the light bulb.

b. The steel wool fibers remain unchanged and the light bulb turns on.

c. The thin fibers have higher electrical resistance than the thick fibers and clumps, so they heat up to a higher temperature, flash, and melt. The thick fibers and clumps, having lower resistance, allow the passage of enough electricity to light the bulb.

4. The light bulb filament belongs at the top of this resistance list for two reasons: First, using direct observation, it appears to be very thin, even when compared to the fine steel wool fibers. Second, and most convincing, it releases a great deal of energy, mostly in the form of light, when electricity passes through it.

1. bulb filament
2. thinner steel strand
3. thicker steel strand
4. foil ribbon

SURPRISE CIRCUITS

1 Diagram this circuit:

1 bulb
1 cell } *in*
1 switch } *parallel*

2 Now build the circuit. When you push the switch on, what is the surprise?

3 Explain how your circuit works.

USE THE IDEA OF RESISTANCE IN YOUR ANSWER.

4 Build this other circuit.

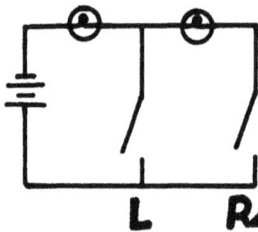

CALL THE LEFT SWITCH "L" AND THE RIGHT SWITCH "R"

L   R

5 Use diagrams and arrows to show how electrons move through this circuit when you turn on. . .

.............**a. SWITCH L ONLY**.............

.............**b. SWITCH R ONLY**.............

.............**c. BOTH SWITCHES L and R**

6 What is the surprise when both switches are turned on? Use the idea of resistance to explain what you see.

Objective

To appreciate that the flow of electricity decreases with increased resistance.

Introduction

None required.

Lesson Notes

2. The bulb holder was altered somewhat in the previous 2 activities. (A straightened paper clip was pushed underneath the penny, replacing the foil ribbon.) Students should now put things back to normal, replacing the paper clip with the foil ribbon, before building this circuit.

Caution students not to leave the switch pushed down for extended periods of time. This creates a short circuit that will rapidly drain the dry cells of energy.

Extension

Divide your class into lab groups with four students each. Ask each group to diagram a supercircuit using their bulbs, switches, and cells in any combination they choose.

When designing a circuit, students should not expose any bulb to more than 4.5 volts (3 cells in series). Energy draining short circuits must also be avoided. Require each lab group to fully diagram their circuit, then get your approval before they build it. If you discover a short, or the possibility of burning out a bulb, ask students to make appropriate design changes.

Each group should then build the circuit they have designed, and write a report describing its properties. If time permits, these circuits could be demonstrated before the entire class.

Here is one example of a supercircuit. There are many possibilities.

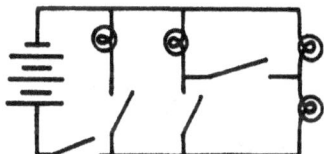

Be prepared for lots of questions, some that you and your students may not be able to answer. Knowing it all, of course, is not the point. The point is to allow your students to experience science as a process of inquiry: to question, argue, hypothesize, experiment, and observe; to experience the satisfaction of being right; to wonder why they are wrong.

Check Point

1.

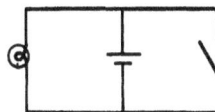

2. The switch works backwards. When you push the switch on, the light goes off. When you lift the switch back off, the light goes on.

3. When you turn the switch on, an alternate low resistance path is provided. Most of the electrons follow this path of least resistance, by-passing the bulb.

It is technically incorrect to say that *all* the electrons follow the path of least resistance through the closed switch. Ohm's law predicts that electricity flows through *both* the bulb and the switch inversely proportional to their respective resistances. Since the resistance of the switch relative to the bulb is very low, most (but not all) electrons travel this path of least resistance across the switch.

To illustrate this idea further, try building this circuit.

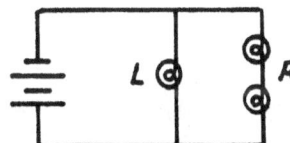

Most of the electrons flow through L because that is the path of least resistance. But the bulbs through R still shine feebly. Because R has twice the resistance, Ohm's law predicts that it receives half the electricity that L receives. Electricity follows the path of least resistance, but it's not an all or nothing principle.

5a. **SWITCH L ONLY**

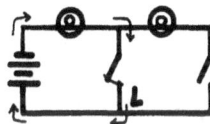

b. **SWITCH R ONLY**

c. **BOTH SWITCHES L and R**

6. When both switches are turned on, only one bulb lights. Most of the electrons follow path L through just one bulb because it is the path of least resistance.

NAME: CLASS:

BUILD A FUSE

1 Cut a piece of paper about as big as this box.

2 Pull off a single strand of steel wool and tape each end to this piece of paper.

3 Clamp 2 coins on the steel wool with clothespins. Keep the space between them very small.

VERY SMALL GAP

STEEL WOOL

4 Clamp a foil ribbon over each coin. Keep these coins so close that they almost touch.

NOW IT'S A FUSE.

5 Put this fuse in series with 2 dry cells and a bulb: the bulb should shine brightly.

ALL the electrons flow through the steel wool in this tiny gap!

IF THE FUSE MELTS, MAKE ANOTHER. USE A LITTLE WIDER GAP.

6 While the bulb is shining brightly, by-pass it with another ribbon. Keep your eye on the fuse!

BY-PASS

IF THE FUSE WON'T MELT, MAKE ONE WITH A NARROWER GAP.

7 Electric wires in a house can get too *hot* if. . .

...the appliances are by-passed (*short circuit*).

...too many appliances draw electricity off the same line (*overload*).

a. How does a fuse protect a house from fire?

b. A fuse should be neither too weak nor too strong. Explain.

Objective

To understand how fuses work to protect circuits from shorts and overloads.

Introduction

Pass a car fuse among your students. (Look for them under the car's dashboard near the steering column, or under the hood.) Draw an enlarged picture of this fuse on your blackboard. Point out how the fuse's conducting strip narrows in the middle, creating a weak point that can easily melt if there is a short circuit or overload.

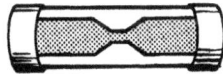

Lesson Notes

2. Begin with a fiber of steel wool about as thick as a hair. Fibers that are much thicker than this may not melt. Tape this fiber at the *ends* of the paper only. The middle must remain free of tape.

3-4. Be sure both coins touch the steel wool. For the electricity to pass through the fuse, it must flow from one coin, across the steel wool, to the other coin.

It is crucial to keep this gap narrow, about 1 millimeter across. Students commonly make it much wider than it should be.

5-6. The bulb must shine brightly in step 5. Otherwise there will be an insufficient surge of electricity in step 6 to melt the fuse when the bulb is by-passed.

Your students may have to try several times before making a fuse that works. The trick is to make it strong enough to sustain the light bulb, but weak enough to melt when the light shorts out.

Students can control the strength of their fuse by adjusting the width of the gap across the two coins:

Gap is too wide. Bulb will not shine brightly, nor will the fuse melt.

Gap is still too wide. Bulb shines brightly, but the fuse won't melt when it shorts out.

Gap is just right. Bulb shines brightly and the fuse melts when it is shorted out.

Gap is too narrow. The fuse melts immediately, before the bulb is shorted out.

Extension

Tape a strand of steel wool to a balloon like this.

Lay 2 foil ribbons, end to end, along this strand so the ends *almost* touch. Place tape across this gap.

Connect these ribbons to an open switch and 2 dry cells.

Discuss what happens when you close the switch. (Electricity flows through the high resistance strand of thin steel wool. It heats up, melts the skin of the balloon and causes it to pop.)

Check Point

7a. A fuse is the weakest link in a circuit. If too much electricity flows through the line because of a short circuit or overload, the fuse is the first to melt and break the circuit, before other wires get hot enough to catch the house on fire.

7b. If the fuse is too weak, it will melt whenever electricity flows through the circuit, whether or not there is an overload or short in the circuit. If the fuse is too strong, it won't melt in time to protect other wires from overheating.

K. MAGNETS

K-1 is it magnetic?
K-2 name that pole
K-3 invisible gears

K-4 up in the air
K-5 how strong?

Magnets are a natural wonder. They sometimes attract and sometimes repel; they attract some materials but not others. Spin one magnet and watch another rotate nearby at the very same speed. Make a paper clip float in mid-air.

In these activities your students will discover that a magnet's force can pass unaltered through many materials — paper, aluminum foil, the human hand —

anything non-magnetic. They'll graph how this force diminishes with increased distance. Paper clips measure the strength of this force in quantitative terms. Layers of tape separate these paper clips from the magnet by uniform distances.

No problem motivating students to study about magnets. Students are attracted to them quite naturally.

——————— EVALUATION ———————

Each question evaluates a single activity from MAGNETS as numbered. Use any combination to frame a formal exam or an informal review: Copy these questions on your blackboard, construct your own ditto master, or photocopy the questions while masking out the rest of the page. Evaluate in ways that suit your own teaching style, enabling your students to learn and enjoy science.

Questions

K-1
A metal shop wishes to recycle the copper and iron shavings falling on its floor. How should they separate this mixture of metals they sweep up at the end of each day?

K-2
Pole B on magnet AB attracts pole Y on magnet XY. When you hang magnet XY from a thread, pole Y points to the Earth's north. Name each pole on the line provided.

K-3
Three magnets hang so they almost "grab" each other.

a. One pole is marked north (N). Label the other five poles.

b. Could you turn magnet C by turning magnet A? Explain.

K-4
A compass needle is actually a tiny magnet. It freely rotates in the earth's magnetic field coming to rest in a north-south line. As a compass maker, would you surround your compass needles with iron or aluminum? Explain.

K-5
What does this graph tell you about a magnet?

Answers

K-1
Pass a magnet through the pile of metal shavings. The iron will stick to the magnet, leaving the copper behind.

K-2
Y is north by definition.
B is south because it attracts Y.
X is south because it's opposite Y.
A is north because it's opposite B.

K-3 a.

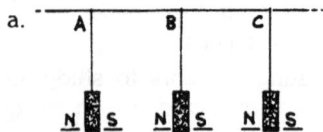

b. Yes. A turns B, which turns C.

K-4
Encase the compass needles in aluminum. Earth's magnetic field will pass through aluminum relatively undisturbed, but not through iron. Iron would distort the earth's magnetic field, causing the needle to stray from the north-south line.

K-5
The attractive force of a magnet on any magnetic object decreases as you separate one from the other, and increases as you bring the objects back together. (According to Coulomb's law, the strength of a magnetic field decreases inversely with the square of the distance.)

SEQUENCING

MAGNETS is an introduction to simple magnets and how they work. Both L and M follow logically, though **K** is not a mandatory prerequisite. If you decide to skip **K** or do L first, be sure to label the poles of all magnets per instructions in K-2 first.

Related Activities: **K**---L—M

MATERIALS

Here is everything your students will use for the next 5 activities on MAGNETS. Materials printed in normal type are part of the core 15-things-in-a-box inventory that support all 100 activities. Materials printed in *italics* are additional local materials that you provide or ask your students to bring from home. Pencil and paper are already assumed and therefore unlisted. Each item is numbered with the activity where it is first used.

(K-1) Small ceramic magnets.

ACTUAL SIZES:

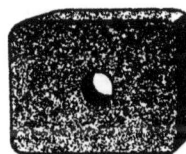

(K-1) Objects to test for magnetic attraction. *See "Preparation" in teaching notes K-1.*
(K-1) Thread.
(K-1) Steel pins.
(K-1) Aluminum foil.
(K-1) Paper clips.
(K-1) Copper wire.
(K-2) Masking tape.
(K-2) Scissors.
(K-4) Clothespins.
(K-4) *Medium-sized cans.*
(K-4) *Scratch paper.*

FURTHER STUDY

Use problems like these plus "extension" ideas in MAGNETS to lead your students beyond worksheet activity into original research and investigation. Each discovery leads to more questions, deeper questions, better questions than these. Answering them is what good science is all about.

Find out all you can about lodestone. Write a story telling how you think this substance was recognized for the first time and how it might have affected the life of the person who discovered it. Be creative.

Do the Earth's north and south magnetic poles remain in a fixed place? Write a report about magnetic drift.

What role do magnets play in producing electricity? Read about generators and write a report.

NAME: _____   CLASS: _____

IS IT MAGNETIC?

1 List *magnetic* things in the *left* table; *nonmagnetic* things in the *right* table.

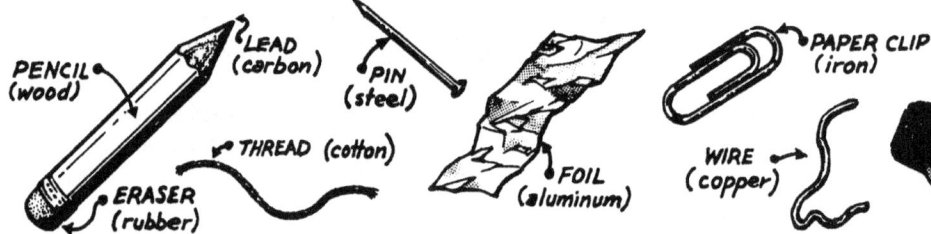

PENCIL (wood) LEAD (carbon) PIN (steel) THREAD (cotton) ERASER (rubber) FOIL (aluminum) WIRE (copper) PAPER CLIP (iron)

MAGNETIC: attracted to a magnet		NONMAGNETIC: not attracted to a magnet	
OBJECT:	MADE FROM:	OBJECT:	MADE FROM:

2 Check all materials that are made from metal.

...NETIC: attracted to a magnet

MADE FROM: ✔

...steel ✔

NON... OBJECT:

window   glass
shoe laces   nylon
...can   aluminum ✔
   plastic

3 Are *all* metals magnetic? Explain.

4 Are all *nonmetals* *nonmagnetic*? Explain.

5 What kinds of material always seem to be magnetic?

Objective

To recognize that only a few metals, like iron and steel, are attracted by a magnet.

Preparation

Gather a collection of objects in a box, both metal and non-metal, to test for magnetic attraction. Be sure to include all the materials pictured on the worksheet, plus additional items at hand. Include different kinds of metals — iron, steel, copper, brass, aluminum, etc.

Lesson Notes

1-2. These 6 illustrated objects suggest only a starting point. If interest remains high encourage students to extend both tables.

Be prepared for lots of questions: What do you call this? How do you spell it? What's it made from? In general, any magnetic object found in your classroom will likely contain iron. Nickel and cobalt are also magnetic elements, but these are relatively rare.

Your students will also find magnetic things made from steel. Steel is an alloy of iron. It is usually mixed with carbon (sometimes with other metals) to improve strength and hardness. Iron is seldom encountered alone, as a chemically pure element.

Whether you say a paper clip is make from iron or steel is a question of semantics: how much carbon should iron have before you call it steel? One way to decide is to consider the metal's hardness. Greater amounts of carbon are alloyed with iron to produce harder steel. A relatively soft iron paper clip might be distinguished from a hardened steel pin on this basis. But this distinction is of little consequence; allow your students to interchange the terms iron and steel as they please.

Other questions that may arise: A tin can is magnetic because it is made from tin-plated iron (or steel). Ceramic magnets are magnetic not because they are made from a special sort of clay. They are made from iron oxide. Nickel is a magnetic element, although the US nickel coin (an alloy of 25% nickel and 75% copper) shows no visible attraction to a magnet. Canadian nickels, however, are magnetic.

Check Point

1-2. Students may list materials other than these:

MAGNETIC:

OBJECT	MADE FROM . . .
pin	steel ✓
paper clip	iron ✓
nail	iron ✓
scissors	steel ✓
staples	iron ✓
steel wool	steel ✓

NONMAGNETIC:

OBJECT	MADE FROM . . .
pencil's shaft	wood
foil	aluminum ✓
pencil's lead	carbon
thread	cotton
eraser	rubber
wire	copper ✓

3. No. Some metals are magnetic (iron and steel); other metals are not (copper and aluminum).

4. Yes. All the non-metals tested were non-magnetic.

5. Iron and steel.

NAME THAT POLE

1 Bring 2 magnets close together.

a. What happens when the magnetic poles *attract?*

b. What happens when the magnetic poles *repel?*

2 Cover both poles on each magnet with masking tape.

TAPE

TAPE

3 Draw large *circles* on two poles that repel.

CIRCLES

Draw large *squares* on the other sides.

SQUARES

4 Use your magnets to complete this table.

like poles	○ to ○	*repel*
unlike poles	○ to □	
	□ to □	
	□ to ○	

5 Tape a loop of thread to your table. Make it about as long this paper. . .

LENGTH OF PAPER

Then hook 2 paper clips into your loop. . .

Then hang one of your magnets between the paper clips. Allow it to come to rest.

6

N or S?

a. The circle or square? _____ points north. Label it **"N"**.

b. The circle or square? _____ points south. Label it **"S"**.

Label both poles on your other magnet.

7 Make sure you labeled all poles correctly:

COMPLETE THIS CHECKLIST!

When you hang either magnet from the loop:

a. ☐ N faces earth's North.

b. ☐ S faces earth's South.

When you bring both magnets together:

c. ☐ S repels S.

d. ☐ N repels N.

e. ☐ S attracts N.

Objective

To identify, then label, the north and south poles on unmarked magnets by using the Earth's magnetic field as a reference.

Preparation

To successfully complete this activity, your students must know how their classroom is oriented with respect to Earth's magnetic north-south axis. If directions are not obvious to all, you'll need to make a sign. Write a large "N" on one sheet of paper and a large "S" on another. Hang these papers on the wall indicating north and south relative to the center of your room. For an even better visual representation, tape the "N" and "S" to a meter stick or long piece of wood, and hang it at the proper orientation from your ceiling.

Magnetic north and geographic north don't coincide in most areas. (This concept is treated as an extension in activity L-3.) For now, orient your signs to magnetic north and south, not true geographic north and south. A magnet taped to a thread will indicate the proper alignment.

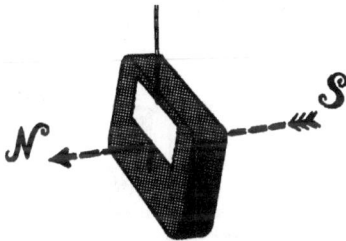

If you are unsure which end is north in this alignment, consult a compass or observe the position of the sun in your sky at high noon.

Lesson Notes

The two areas of greatest force on a magnet are called its north and south poles. These poles are located on the faces of each magnet, one on each side. In this activity, your students will decide by experiment which of these poles is north and which is south.

5-6. By definition, the north-seeking pole on a magnet is called "north"; the south-seeking pole is called "south". Students may question this definition. If like poles repel, how can they point to each other? In reality, the Earth's north magnetic pole is really a south pole. And the Earth's south magnetic pole is really a north pole. This has long been a source of great confusion.

But the confusion would be even worse if mapmakers and geographers located magnetic north and geographic north half a world apart. So the convention stands. Encyclopedias and atlases show magnetic north next to geographic north, even though it really isn't. Earth is a mislabeled magnet!

Check Point

Before students come to you for a check point, make your own "checking magnet." Label it with a bold clear "N" on one side and "S" on the other. You can do this by following the steps in this activity sheet. Or simply hang the magnet from some thread with the poles pointing out; it will always come to rest so its north pole faces Earth's north. However you do it, do it *yourself* so you *know* the magnet is labeled correctly. Use this checking magnet to verify that *all* student magnets have been labeled correctly. If some have been labeled wrong, great confusion will result in later activities.

When your students ask for a teacher check, hold your checking magnet in one hand, and one of your student's two magnets in the other. Ask, "Will these poles attract or repel?" Let your student predict, then test each hypothesis. This will solidify the concept of like and unlike poles, providing good closure to the lesson.

At the end of the day, when all magnets are labeled and handed in, there is one final way to check for pole accuracy. Place all the magnets together in one long row. All north poles (and all south poles) will naturally face the same way. Any anomalies are easy to spot and should be relabeled.

N's FACE THE SAME DIRECTION

1a. They come together and remain attached.
1b. They push away and remain apart.

4.

like poles	○ to ○	repel
unlike poles	○ to □	attract
like poles	□ to □	repel
unlike poles	□ to ○	attract

6. Varied answers. On some magnets the square will be north. On others, the circle will be north.

7. a. yes
    b. yes
    c. yes
    d. yes
    e. yes

INVISIBLE GEARS

1   Tie a loop of thread about as long as this paper. . .

Tape one of your magnets inside this loop.

TAPE

Tape this loop to your table so the magnet hangs over the edge.

TAPE

POLES

THE POLES FACE SIDEWAYS, NOT UP AND DOWN

2   Hang another magnet level with the first. Keep them as close as you can without the 2 magnets grabbing each other.

3   Rest one magnet on the table. Spin the other around and around until you wind up the string.

4   Allow both magnets to rest side by side. Then let them go.

LET THEM HANG WITHOUT SWINGING...

...THEN LET BOTH GO TOGETHER.

If they don't start spinning, give either magnet a gentle push.

Write your observations.

5

a. As one magnet winds down, the other magnet ....

b. As one magnet loses energy, the other magnet ....

c. The magnets turn so that............................... poles always face each other.

d. What is the "invisible gear" that links these 2 magnets together?

....

e. Can this invisible gear pass through your hand? Explain how you know.

....

....

....

Objective

To observe interactions between rotating magnetic fields.

Introduction

None required.

Lesson Notes

1. To form a loop as long as this worksheet, you need to start with a piece of thread at least twice this length. Some students may overlook this, and end up with only half the desired loop length.

The magnet must hang in a vertical position, with the poles pointing out along the horizontal. It's best if you tape this magnet to the thread along its top edge. If it still tilts, you can make minor corrections by pulling one side of the loop or the other a small distance through the tape.

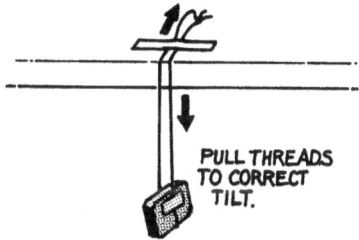

PULL THREADS TO CORRECT TILT.

2. If your table legs are made of iron, hang these magnets near the center edge of the table, as far from the iron legs as possible. If the table top itself contains iron, make the loops long enough so attraction between the magnet and the table top is reduced to a minimum. The distance between the magnets is critical. If they are too close, the magnets won't be able to spin without grabbing each other. If they are too far apart, the magnetic linkage effect in step 4 won't be as dramatic.

3. The fastest way to wind the magnet on its thread is to give it a quick twist with the wrist or a sharp snap of the finger, then let it spin freely. When it comes to rest, spin it again in the same direction.

4. The secret here is to cradle both magnets between your fingers so they hang almost motionless, side by side. Then let them go. If the magnet is wound tightly enough, both will begin to spin on their own. If not, give either magnet a gentle nudge.

Your students may not successfully observe this magnetic linkage the first time. Encourage them to fine-tune the system, adjusting the spacing between the magnets until they are successful. Such trial-and-error tinkering is at the heart of scientific inquiry.

Check Point

4. The magnets appear linked together. As one starts turning, speeds up, slows down, stops, begins turning in the other direction — the other magnet matches these varied motions turn for turn.

5 a. winds up
   b. gains energy
   c. unlike
   d. magnetic attraction
   e. Yes. Put your hand between the magnets and they still turn like gears.

UP IN THE AIR

1 Clamp a magnet in a clothespin. Tape it to the bottom of a can like this.

Tape only the BOTTOM WING.

2 Tie a paper clip to some thread. Touch it to the bottom of the magnet so the thread hangs down.

3 Tape the thread to the table, but leave the end free.

LEAVE END LOOSE

4 Now pull the thread so the paper clip hangs "up in the air" away from the magnet. Keep the space between as wide as you can.

5 Pass these materials through the magnetic field that holds the paper clip.

Write your observations.

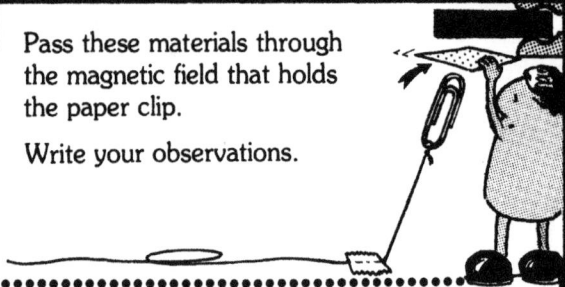

a. Paper:

b. Pin:

c. Paper Clip:

d. Aluminum Foil:

6 **a.** *Predict* if copper wire will disturb the magnetic field. Explain your reasoning.

b. Test your prediction. ☐ **correct** ☐ **wrong**

7 **a.** *Predict* if iron scissors will disturb the magnetic field. Explain your reasoning.

b. Test your prediction. ☐ **correct** ☐ **wrong**

Objective

To observe that a magnetic field passes unchanged through solid objects unless they are magnetic.

Introduction

None required.

Lesson Notes

4. You can also hang the paper clip up in the air by sliding the can slightly away.

5. The force that surrounds a magnet is called its "field". This field will pass right through non-magnetic materials like the paper and aluminum foil, with no visible effect on the paper clip. But the field is significantly altered by magnetic objects like the pin and paper clip. These iron objects become magnets themselves, when placed near the permanent magnet. For a more detailed examination of interacting magnetic fields, see activities L-5 and L-6.

Students who claim that paper or aluminum foil also cause the paper clip to wiggle or fall haven't exercised enough care in passing these test objects through the field. Younger students, in particular, may not be coordinated enough to avoid bumping the suspended paper clip.

6-7. The important generalization to make is this: only magnetic objects interact with the field and cause the paper clip to wiggle or fall. Thus iron scissors will disturb the field, while copper wire will not.

This classification of substances as magnetic or non-magnetic, while suitable for an elementary understanding of magnetism, should nevertheless be recognized as an over-simplification. Non-magnetic substances (implicitly defined in these activities as materials that are not visibly attracted to a magnet) *do* in fact interact with other magnets. But this interaction is too weak to observe directly.

Whenever electrons move, they produce an associated magnetic field (see M-1). In a sense, then, anything with electrons (all matter) has magnetic properties.

Most substances are diamagnetic. Their electrons weakly oppose external magnetic fields at either pole. A few substances, because of unpaired electrons, are paramagnetic. They weakly increase external magnetic fields at either pole. Only three elements — iron, nickel, and cobalt — are ferromagnetic. Their special electron configurations enable these elements to remain permanently magnetized.

In these activities, ferromagnetic substances are simply called magnetic. The other forms of weak magnetism, being too subtle to observe directly and too complicated for elementary study, are not considered.

Check Point

5a. Paper: the paper clip remains undisturbed.
 b. Steel Pin: the paper clip wobbles and sometimes falls.
 c. Paper Clip: the other paper clip gets pushed aside and falls.
 d. Aluminum Foil: the paper clip remains undisturbed.

6a. Non-magnetic materials (the paper and aluminum foil) did not disturb the field. Since copper wire is also non-magnetic, it shouldn't disturb the field either.
 b. Prediction is correct.

7a. Magnetic objects made from iron (the pin and paper clip) disturbed the field. Because scissors are also made from iron, they should disturb the field as well.
 b. Prediction is correct.

HOW STRONG?

1 Clamp a magnet in a clothespin and tape it to the bottom of a can. Pull apart a paper clip (just a little) and hang it underneath.

MAKE A HOOK →

TAPE BOTTOM WING

2 Now add paper clips. Write the total number (support clip included) that the magnet can hold next to the "0" in the table below.

THIS MAKES **4**.

3 Cut a 2 cm piece of masking tape and stick it under the magnet.

|← 2 cm →|

Find how many clips your magnet now holds.

Write your answers in the table next to the "1"

4 Add more layers of tape as listed in the table. Record your results.

STACK THE LAYERS NEATLY

5 Make a graph. Draw the best smooth curve you can through the points.

CUTOUT

K-5

DISTANCE (TAPE LAYERS)	STRENGTH (PAPER CLIPS)
0	
1	
2	
4	
8	
12	
20	
30	
40	
60	

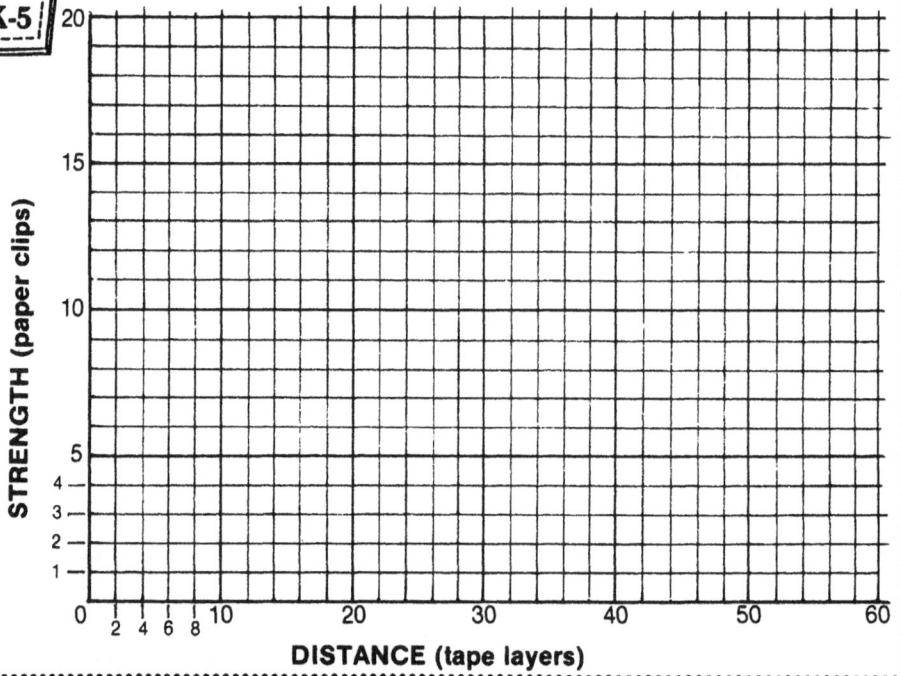

STRENGTH (paper clips) vs DISTANCE (tape layers)

Objective

To graph how the strength of a magnet decreases with increased distance from the magnet.

Introduction

Recall from the previous activity how a magnet attracts a paper clip at a distance. This attraction was strong enough to hold the paper clip up in the air as long as the distance to the magnet remained close. Pull the paper clip too far from the magnet and down it falls.

This implies that the strength of a magnetic field decreases as the distance from the magnet increases. The purpose of this activity is to determine how rapidly a magnet's strength drops off with increased distance. Strength is measured by how many paper clips a magnet can hold. Distance is measured in layers of tape.

Each piece of tape added to the magnet removes the paper clip from the surface of the magnet by another small unit of distance. The tape itself does not interfere with the magnetic field because it is non-magnetic.

Lesson Notes

Allow plenty of time for this activity — perhaps an hour. If your students need to continue their experiments-in-progress the next day, be sure they continue with the *same* magnets.

1-3. Notice that only the bottom half of the clothespin is taped to the inverted can. The top half remains free so you can easily remove the magnet from the jaws of the clothespin.

The support paper clip is bent just far enough to allow other paper clips to slip through the gap in the middle. Its shape prevents the attached clips from scattering when they fall to the table. And fall they do — again and again — as students try to add just one more clip to the cluster under the magnet.

There is already one piece of masking tape stuck to the surface of the magnet. Your students can remove this if they wish to truly begin with "zero layers." (They should put it back afterwards.) Or they can think of this tape as an integral part of the magnet itself, so that magnet *and* tape form the zero-layer starting point.

4. To maintain a flat even surface, students must cut *uniform* pieces of tape and add them *evenly* to the bottom of the magnet. Otherwise a rounded mound results.

Don't cut the tape too long. If pieces extend under the jaws of the clothespin, the magnet will tilt.

Some students may fail to appreciate that the number of tape layers in the data table represents an accumulated total: 1 new layer makes a total of 2, 2 new layers make a total of 4, 4 new layers make a total of 8, and so on.

To avoid confusion, ask students to cut squares of masking tape in advance, number them from 1 to 60, and store them on a glossy surface. (A manila folder is ideal).

Students then apply these squares of tape to their magnets in numerical order, keeping an accurate running total. If they run out of time one day, it is easy to pick up where they left off on the next.

5. Random experimental error will scatter the data points somewhat. Your students should understand that it is *not* essential to connect *every* data point on the graph with a line. Instead, they should draw the best possible smooth curve to fit the overall trend suggested by the scattered points.

Check Point

2-5.

DISTANCE (TAPE LAYERS)	STRENGTH (PAPER CLIPS)
0	12
1	12
2	11
4	10
8	9
12	7
20	5
30	3
40	2
60	1

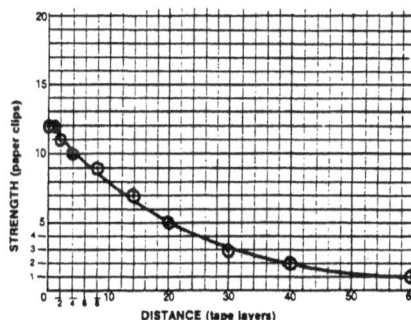

L. MAGNETIC FIELDS

These activities turn on pins, common steel pins. Touch one to a magnet, and the pin itself becomes a magnet. Pivot this pin in the Earth's magnetic field and it always points in the same direction when it comes to rest, just like a compass. Use this pin to plot compass directions and make maps.

Tie this magnetized pin to thread and let it hang near a larger magnet. The pin continuously aligns itself to the magnetic field of the larger magnet when moved from place to place. By drawing the pin's alignment at each position, students can map the shape of the entire field.

Interacting magnetic fields are especially interesting to study, both those that attract and those that repel. The simple magnetized straight pin again reveals all — a remarkable scientific tool.

——— EVALUATION ———

Each question evaluates a single activity from MAGNETIC FIELDS as numbered. Use any combination to frame a formal exam or an informal review: Copy these questions on your blackboard, construct your own ditto master, or photocopy the questions while masking out the rest of the page. Evaluate in ways that suit your own teaching style, enabling your students to learn and enjoy science.

Questions

L-1
A pin is glued to a small chip of wood so it floats on a bowl of water.

Explain how to make a compass so the pin head points north.

L-2
Make a small dot in the center of a clean sheet of paper, then tape the paper to your table. Use your ruler and hairline compass to solve this letter puzzle.

dot • → NW / 5 cm
dot • → N / 7 cm
dot • → E / 7 cm
dot • → SE / 5 cm

Write letter here.

L-3
Label the 4 opposite corners in your room. Find these directions using your pin compass.

corner C ——→ corner D

corner D ——→ corner A

L-4
Finish this picture. Use long, smooth lines to represent the lines of force in the magnetic field.

N S

L-5
Label the other 5 poles not shown.

N

L-6
Finish this picture. Draw lines of force to show the shape of the interacting magnetic fields.

S N   N S   N S

Answers

L-1
Magnetize the pin by touching its head to the north pole of a magnet and its point to the south pole of a magnet. Then float it on its wood chip in the bowl of water. The pin will align itself to the Earth's magnetic field. No matter how you turn the bowl, the pin head always points north.

L-2
K

L-3
Answers depend on your choice of reference corners.

L-4
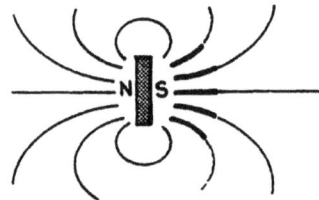

L-5

L-6

SEQUENCING

MAGNETIC FIELDS follows logically from K . It is not a prerequisite as long as you first identify the poles of all your magnets in K-2. **L** is a prerequisite to M because the compass in L-1 is required in M-2.

Related Activities: K---L—M

MATERIALS

Here is everything your students will use for the next 6 activities on MAGNETIC FIELDS. Materials printed in normal type are part of the core 15-things-in-a-box inventory that support all 100 activities. Materials printed in *italics* are additional local materials that you provide or ask your students to bring from home. Pencil and paper are already assumed and therefore unlisted. Each item is numbered with the activity where it is first used.

(L-1) Scissors.
(L-1) Masking tape.
(L-1) Steel pins.
(L-1) Aluminum foil.
(L-1) Magnets with labeled poles.
(L-1) Clothespin.
(L-1) *Medium-sized cans.*
(L-1) Rubber bands.
(L-3) Clear tape.
(L-4) Size-D dry cells.
(L-4) *Scratch paper.*
(L-4) Thread.

FURTHER STUDY

Use problems like these plus "extension" ideas in MAGNETIC FIELDS to lead your students beyond worksheet activity into original research and investigation. Each discovery leads to more questions, deeper questions, better questions than these. Answering them is what good science is all about.

How did sailors navigate before the invention of the compass? What forms of navigation do they use today in ships of iron and steel? Write a report.

Find the magnetic variation (declination) for your area. Where else on Earth is this variation the same as where you live? Copy an isogonic map of the world.

Use a pair of compasses and a protractor to box the compass. Include at least 16 directions in your coordinate system. Let north be 0° and east be 90°.

PIN COMPASS

1 Cut a square of masking tape about this big.

Stick it on your desk or other clean surface.

2 Write the 4 compass directions on your square, like this.

BOLD, CLEAR LETTERS.

N
W E
S

3 Peel up the tape, holding the sticky side up.

Lay 2 pins (point-to-point) on the sticky side directly over the "S" and "N".

LEAVE A SMALL GAP

4 Cover with a small piece of foil.

Trim off the excess around the edge of the square.

FOIL
TAPE

5 Touch the "N" pin head to *south* on a magnet.

Touch the "S" pin head to *north* on a magnet.

6 Rubber-band one wing of a clothespin to the side of a tin can. Stick a pin point up, at the top.

CAREFUL — SHARP!

RUBBER BAND

7 Fold your compass square along its N-S and E-W axes. . .

N-S FOLD:

E-W FOLD:

...so it can balance and turn on the pin.

8 Slowly turn the can around while watching the needle.

WATCH!

Have you made a compass? Explain.

TURN.

★Save your compass. ★

Objective

To build a pin compass and observe how it works.

Introduction

Build this compass yourself before your students try. This will familiarize you with the directions, and provide a model for your students to follow.

Emphasize how the compass assembly rests lightly on *top* of the pin point. (It must never be poked through.) The tape and magnetized pins cannot fall because their center of mass rests below the pivot point. The magnetized pins remain free to turn this way or that, until finally settling into the north-south orientation of Earth's magnetic field.

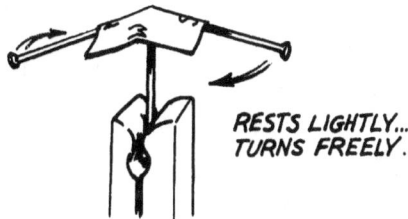

RESTS LIGHTLY... TURNS FREELY.

Lesson Notes

3. These pins don't actually cover the letters "S" and "N". The letters are directly underneath them, it is true, but on the opposite non-sticky side of the tape.

Keep a small distance between these pin points. This is where the compass will eventually pivot in step 7.

5. The pin taped to north on the compass must be touched to south. Likewise, the pin taped to south on the compass must be touched to north. Compasses that point the wrong way in step 7 were incorrectly magnetized here. (Recall from the lesson notes in activity K-2 that Earth is a mislabeled magnet.)

6. Rubber band the inside clothespin wing only. Leave the outside wing unrestricted so you can freely open and close the jaws of the clothespin.

7. Fold the N-S line around the shaft of both pins. Fold the E-W line so the pins "bump heads." This places the pivot (the intersection of these two folds) squarely on center, assuring that the pins will balance level.

Before balancing this assembly on the pin point, bend both fold lines simultaneously about half-way closed. This creates a "tent" shape. The pivot is at the top, with the magnetized pins angled somewhat down. Set this "tent" *gently* on its "support" pole. (Never poke it through.) It should balance easily and turn freely.

8. Students will use their compasses again, in activities L-2, L-3, and M-2. They should write their names on masking tape labels and stick them to the side of the can.

Check Point

8. Yes, the pins move like a compass. The north pin always point north (and the south pin south) no matter how you turn the base of the compass.

LETTER PUZZLES

1 Make a small dot in the center of a clean sheet of paper.

2 Tape the paper to a level surface away from magnets or other iron objects.

3 Cut out the ruler. Use it plus your compass to solve each letter puzzle.

See step 4 for the first puzzle.

4 FIRST LETTER PUZZLE:

DOT —— **NE** 8 cm ——→

DOT —— **S** 8 cm ——→

DOT —— **NW** 8 cm ——→

DRAW **3** STRAIGHT LINES **FROM THE CENTER DOT**

AS YOU SOLVE EACH PUZZLE: **WRITE** THE LETTER IN THE BOX:

5 SECOND LETTER PUZZLE:

Start again with a clean sheet of paper that is taped to your table and make a small dot in the center.

DOT —— **N** 8 cm ——→ **SE** 12 cm ——→

DOT —— **W** 8 cm ——→ **SE** 12 cm ——→

START FROM THE DOT JUST TWICE

6 THIRD LETTER PUZZLE:

Start once more with a clean sheet of paper that is taped to your table and make a small dot in the center.

DOT —— **W** 6 cm ——→ **S** 7 cm ——→ **E** 7 cm ——→ **N** 12 cm ——→

DOT —— **NW** 8 cm ——→

CENTIMETER RULER

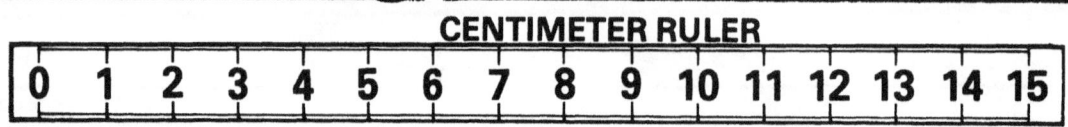

0 1 2 3 4 5 6 7 8 9 10 11 12 13 14 15

Objective

To practice finding directions with a compass.

Introduction

Write a compass direction on your blackboard, SE, for example. Ask your students to find SE on their pin compasses, then point their pencils in the indicated direction. When everyone is in agreement, write another direction on your blackboard for students to point to. Repeat as necessary.

Next group your students into pairs. Each pair spins a pencil on the table and lets it come to rest. Both use a pin compass to decide in what direction the pencil points. They write down answers on separate pieces of paper, then compare.

When students disagree, they can debate among themselves who is right and who is wrong, learning much about compasses and their inherent accuracy. In general, answers should not vary by more than 45°. If one gets NW for example, the other might determine the same direction as N or W and still be within reasonable limits of compass error. When students agree, they should spin their pencil again. Repeat until your students are generally able to agree about compass directions within the limits of compass accuracy.

Lesson Notes

4-6. Read each set of directions from left to right, top to bottom, as if you were reading a book. In step 4, for example, begin at the dot and draw a straight line NE for 8 cm. Begin at the dot again and draw a straight line S for 8 cm, etc. In step 5, notice that 2 sets or arrows are to the right of each dot. This means you begin at the dot and draw 2 lines, end to end, before returning to the dot a second time.

Extension

Organize a treasure hunt. Discuss the following procedures inside your classroom. Then go outside and enjoy the hunt.

a. Cut out 8 small "ground markers" from a piece of paper. Identify each one with your personal initials.

b. Prepare a treasure map outside on the school grounds: Start from any familiar landmark, like a tree or the school steps. Choose a compass bearing, then walk along this line a measured number of paces. Record this information on an index card, then place one of your 8 markers at your feet. (If there is a wind, hold down the marker with a rock.) From this marker, choose a different direction and repeat the process. Use all 8 markers as time allows. Place a "treasure" at the last marker. Your map will look something like this:

Start at the school steps. Walk...

Direction	N	E	SE	E	NNE	W	N	ENE
Number of Paces	20	15	12	7	18	30	10	12

Dig down 5 cm under the last marker to find a coin.

c. When you have completed your map, trade with someone else. Before you start the hunt, find out if your partner has long paces or short paces. Practice walking with the same stride. The grounds will be covered with markers. Be sure to follow markers with the correct initials. Finding each marker may not be as easy as you think. *Both* the map maker and the map follower must give and follow directions accurately in order to locate treasure.

d. When this activity is finished, retrace your own route to pick up all 8 markers with your initials. (You might collect these before students leave the area to insure that all have helped to clean up.)

Check Point

4. Y
5. M
6. R

NAME: CLASS:

Magnetic Fields **L-3**

WHICH WAY?

1 Notice that 4 corners in your room have letters. Use your compass to find each direction *from* your desk.

WRITE EACH DIRECTION

a. **My desk** ....➤ **corner A**

b. **My desk** ....➤ **corner B**

c. **My desk** ....➤ **corner C**

d. **My desk** ....➤ **corner D**

2 Now stand in each corner of your room. Use your compass to find the direction *back to* your desk.

a. **Corner A** ....➤ **my desk**

b. **Corner B** ....➤ **my desk**

c. **Corner C** ....➤ **my desk**

d. **Corner D** ....➤ **my desk**

3 Compare these directions. Do they make sense? Explain.

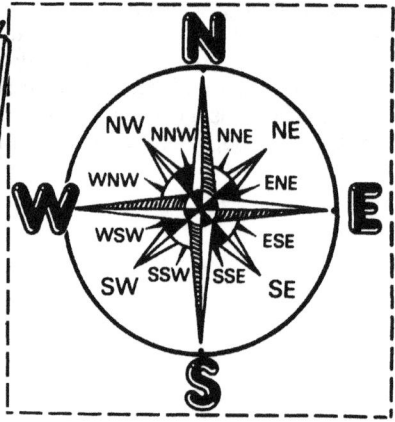

CUT OUTS L-3

CUT ALONG DOTTED LINES

4 Find these directions:

a. **Corner A** ....➤ **Corner B**

b. **Corner C** ....➤ **Corner A**

c. **Corner B** ....➤ **Corner D**

d. **Corner B** ....➤ **Corner C**

e. **This room** ....➤ **School Office**

f. **This room** ....➤ **My home**

5 Draw a *top* view of your room on a full sheet of paper.

➤ Show the location of doors, windows and desks drawn to scale.

➤ Mark the corners A, B, C and D as they appear in your room.

6 Cut out the compass directions. Tape them to your map in the correct position.

Use your map to double check each answer in step 4.

All answers correct? ☐

Copyright © 1988 by TOPS Learning Systems. Reproduction limited to personal classroom use only.

Objective

To practice using a compass to determine directions between reference points in the classroom.

Preparation

Hang a sheet of paper from 4 opposite corners in your classroom. Boldly label these corners A, B, C and D respectively.

Lesson Notes

1, 2, 4. Any sensitive compass requires patience and a steady hand to get the pin to settle down on north. To make accurate readings it may be necessary to rest the compass on a table or the floor.

6. Students should use this map to double check the accuracy of each answer in step 4. (The check box asks students to verify that they have done this.) If answers can't be reconciled, check to see that the compass directions have been taped to the paper in the correct orientation.

Extension

Earth's magnetic poles are near to (but do not coincide with) Earth's geographic poles. From most places on the globe, therefore, the direction to a geographic pole lies on a somewhat different line than the direction to a magnetic pole. The difference between these two lines is called the "angle of variation." (Surveyors call it the "angle of declination.")

You can illustrate this variation in concrete terms. First make a large model compass by drawing coordinates on a full sheet of paper. Place a pencil at the center to represent a compass needle.

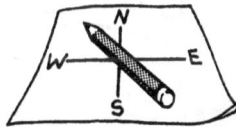

Next, draw a vertical line representing true north on your blackboard. Hang a stick (a meter stick will do) several meters in front of this line to represent magnetic north.

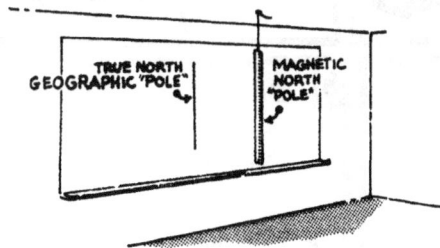

Stand to the left side of your room. Align your model compass to both poles. Aim the pencil at magnetic north, and the north compass coordinate at geographic north. Notice how the compass points perhaps 20° east of true north. That's the angle of variation.

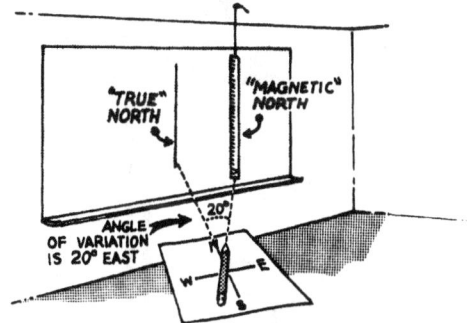

In a similar manner, align your model compass when standing at the center of your room; then to the right. Notice how the size of this variation decreases to zero, then increases again, shifting east to west as you move across the room. This same sort of variation also occurs as you move through different longitudes around the globe.

Check Point

1-2,4. Answers in these steps depend on your choice of the reference corners A, B, C, and D, and their positions relative to magnetic north and south. Once these points are established, student responses should not vary beyond the limitations of compass accuracy. Because all answers have already been double checked, a simple spot check may be sufficient.

3. Going from your desk to any corner is the reverse of going from that corner back to your desk. The directions are 180° opposite. (180° is the ideal case. Because of experimental error, directions may vary by one compass direction, N instead of NE for example.)

5. Check that this map is drawn roughly to scale. Are the tables and aisles proportional to overall room dimensions?

6. Students should state that they have checked all answers in step 4 for internal consistency.

MAP A MAGNETIC FIELD

1 Trace a circle around a dry cell at the center of a full sheet of paper.

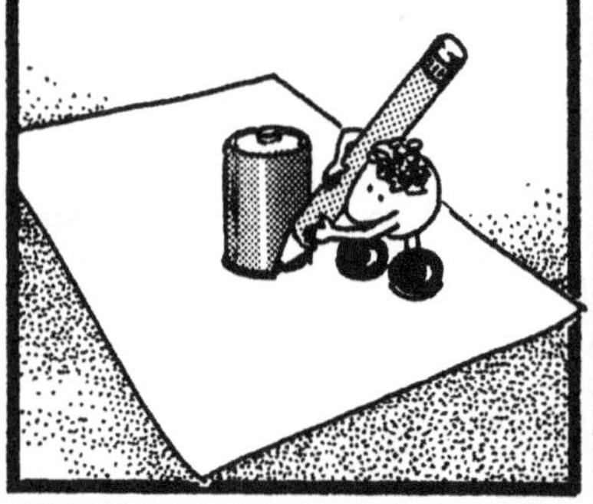

2 Roll a small piece of masking tape, sticky side out.

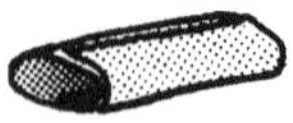

Stick it to the long edge of your magnet.

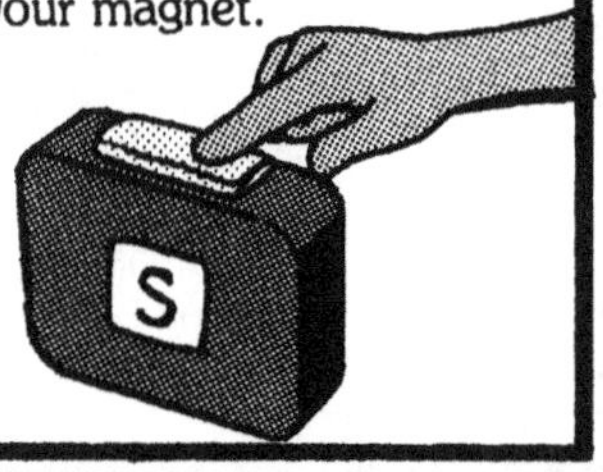

3 Stick the magnet inside the circle:

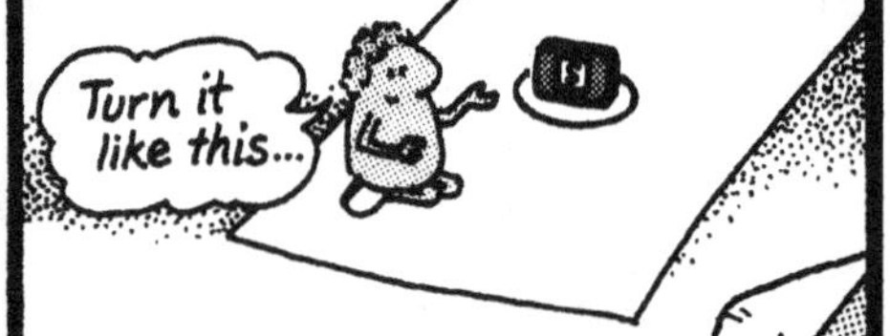

Turn it like this...

Mark South and North in the circle.

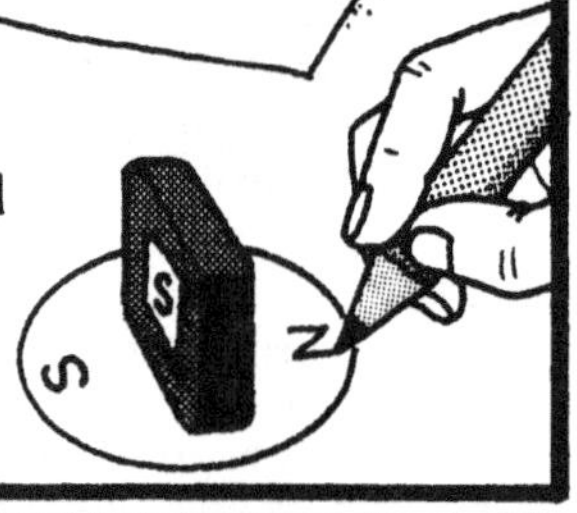

4 Use masking tape to hang a pin from some thread. It should be about this long (10 cm), and hang level.

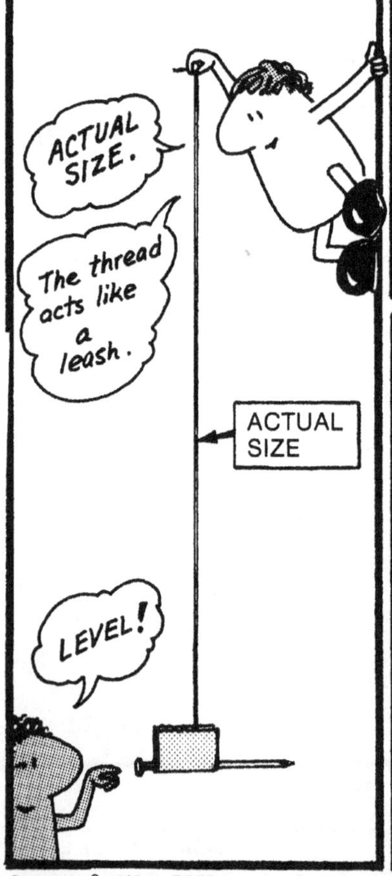

ACTUAL SIZE.

The thread acts like a leash.

ACTUAL SIZE

LEVEL!

5 Touch the head of the pin to south on your magnet.

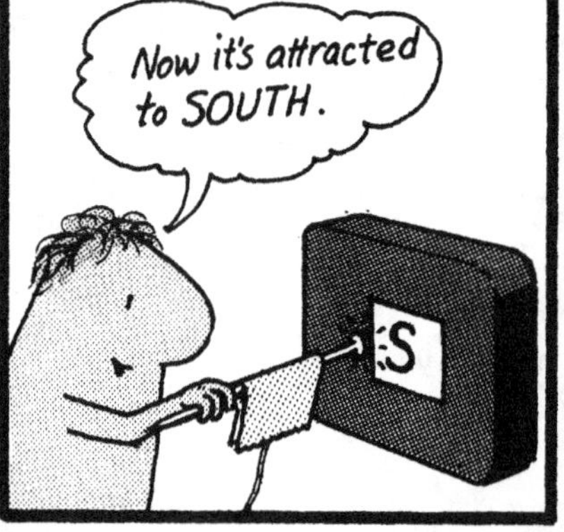

Now it's attracted to SOUTH.

6 Hold the pin by its "leash" at the edge of the circle, near South.

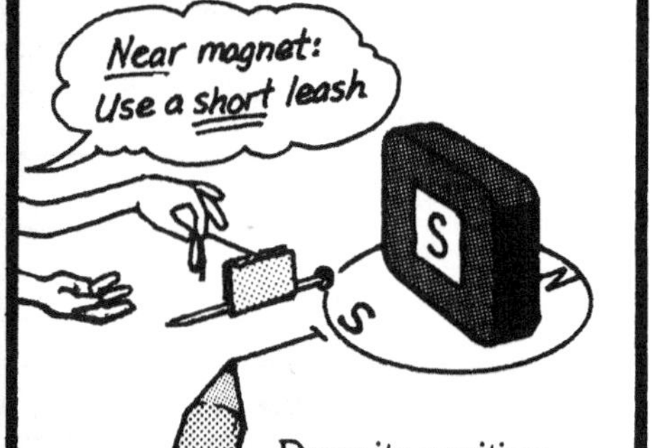

Near magnet: Use a short leash

Draw its position.

7 Draw more pins, head to tail, to make a chain.

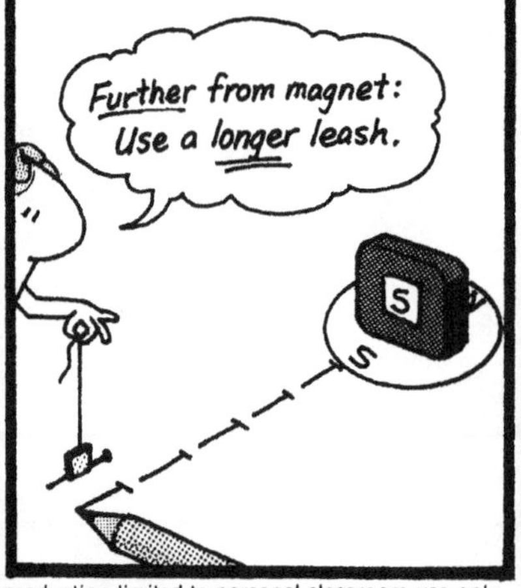

Further from magnet: Use a longer leash.

8 Draw other pin chains, on both sides of the magnet, until you map the entire field.

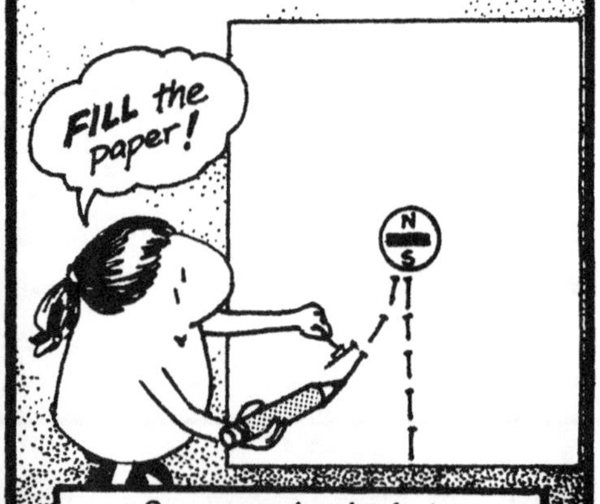

FILL the paper!

Save your leashed pin.

Objective

To map the shape of a magnetic field that surrounds a magnet.

Introduction

Try this activity in advance. Practice drawing pin chains until you are comfortable with the process. Then demonstrate the procedure to your entire class. Emphasize these points as you draw:

a. Begin at the circle. Keep the pin on a short "leash" so it doesn't stick to the magnet.

b. Draw short half-length pins, each time beginning where the previous pin ends. Half-length pins give a better definition of the field than drawing full-length pins.

c. As you map further from the magnet, allow the pin a longer "leash."

d. If the thread gets twisted, let the pin unwind and hang free before you bring it back into the magnetic field.

Draw only one or two pin chains. Don't reveal the crab-like symmetry of the entire field. Allow your students the satisfaction of discovering this on their own.

Lesson Notes

6-8. Students commonly assume that each pin chain radiates straight out from the magnet. They typically complain that the pin turns in the "wrong direction" when it fails to conform to their preconceived notion of what "ought" to happen. The pin, of course, does just what it should do, responding to all the forces acting upon it.

Learning to be objective (seeing things as they really are, instead of how we think they should be) is a skill that requires a lifetime of learning. Speed this process along by pointing out how silly it sounds to blame the pin for misbehaving!

Because the pin head is touched to south, it is attracted to south and repelled away from north. If you think of the pins as arrows (pointing head first), this correctly suggests a convention agreed to among scientists: the magnetic lines of force leave the north pole of a magnet and enter its south pole.

Students should not stop mapping until they have clearly defined the shape of the entire magnetic field. This will require at least 7 chains drawn from each side of the magnet.

Check Point

6-8.

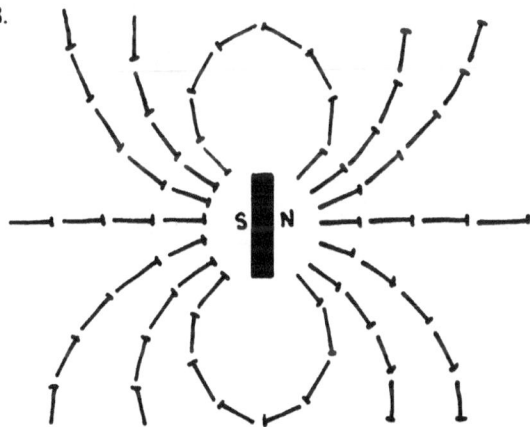

OPPOSITE FIELDS ATTRACT

1   Set up 2 attracting magnets about 15 cm apart.

Set up as before...

Draw 2 circles with a dry cell.

Roll some tape sticky-side-out. Use this roll...

...to tape a magnet into each circle.

N

Mark **N** and **S** in each circle. North must **face** South.

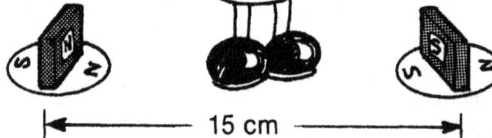

◄——— 15 cm ———►

2   Touch your "leashed" pin head to south...   ...then map the attracting fields with pin chains.

Now it attracts South.

S

Away from magnet, use long leash...

...near magnet, use short leash.

N

N

Draw at least **7** chains.

CENTIMETER RULER

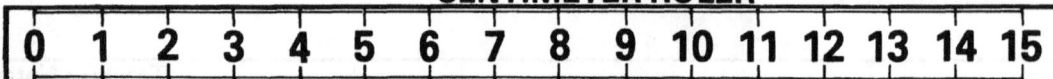

0 1 2 3 4 5 6 7 8 9 10 11 12 13 14 15

Objective

To map the shape of two interacting magnetic fields that attract.

Introduction

None required.

Lesson Notes

1. Students should tape their two magnets so that south faces north. Those who make a mistake here will end up with repelling magnetic fields (the topic of activity L-6), instead of attracting fields.

2. Think of the pins as arrows that point head first. The lines of force move away from north and toward south, assuming a shape that resembles an onion.

If the magnets are correctly positioned but the lines of force move in the wrong direction, the pin has been magnetized backwards. It should be remagnetised by touching the pin head to south as directed.

Extension

Magnetic fields can also be mapped using steel wool. Simply place one or more magnets on a clean sheet of paper in any desired configuration. Turn each magnet so the poles, as usual, point out. Then rub a piece of steel wool vigorously back and forth upon itself so that steel wool dust settles over the entire paper.

STEEL WOOL MAP

If available, you can also sprinkle iron filings over the magnets. This is the traditional way to map a magnetic field.

While this kind of mapping is relatively easy to do, and provides great detail, it is messy and you risk getting slivers in your fingers unless you handle the steel wool very carefully. If you choose to make this extension a class activity instead of a teacher demonstration, consider supplying gloves. Afterwards, the magnets should have their accumulated iron porcupine bristles removed. An easy way to accomplish this is to apply tape over the steel fibers, then peel it back off.

Check Point

2.

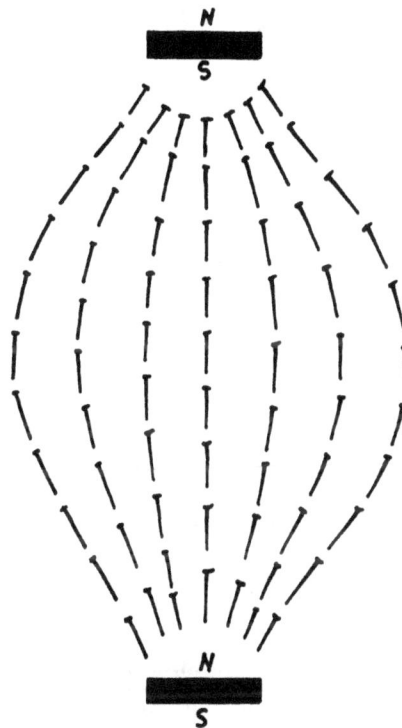

LIKE FIELDS REPEL

1 *Predict* the shape of 2 magnetic fields that repel.

South repels south...

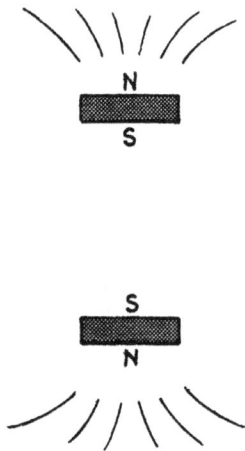

|S ? S|

2 Test your prediction by mapping the field. Set up *repelling* magnets about 15 cm apart, as before.

South must face south.

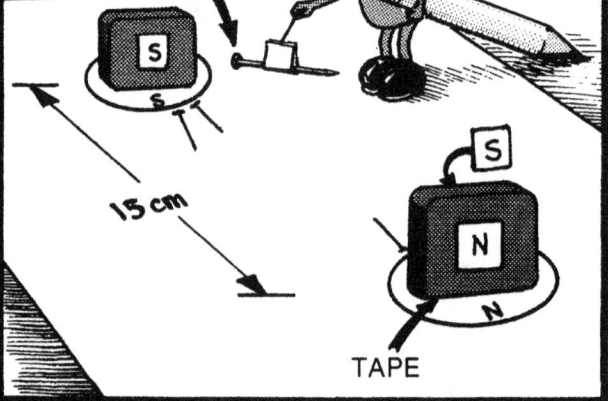

FIRST TOUCH PIN HEAD TO SOUTH

S

S

15 cm

S

N

N

TAPE

N
S

S
N

3 Draw how you think these magnetic fields will interact.

a.

S
N

N
S

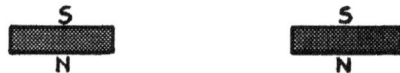

Explain your reasoning.

b.

S
N

S
N

Explain your reasoning.

Objective

To map the shape of two interacting magnetic fields that repel.

Introduction

None required.

Lesson Notes

1. Encourage thoughtful predictions. If the lines of force repel, they should in some manner push each other away. Accept any shape that in some way suggests repulsion.

2. Students should again tape two magnets to their activity sheets, this time with south facing south so that both fields repel. The lines of force appear to clash in this configuration. Fields deflect each other in a mutual repulsion pattern.

The diamond-shaped area in the middle represents a battle ground of sorts. Here opposing magnetic fields clash head-on and compete for dominance. To thoroughly map this area (make the diamond as small as possible), direct pin chains toward the exact center — the point where opposing fields meet in a perfectly balanced stand-off.

As you approach this central point the magnetized pin turns wildly through 180°. Back and forth, around and around it goes. Shifting the pin even slightly brings it into the influence of first one field and then the opposing field.

An easier way to define this difficult central area is to use steel wool. See the extension activity in lesson notes L-5.

Extension

If interest and enthusiasm remains high, here are additional magnetic field configurations to explore. Encourage students to first predict the shape of interacting fields before mapping them with the magnetized pin.

a.   b.

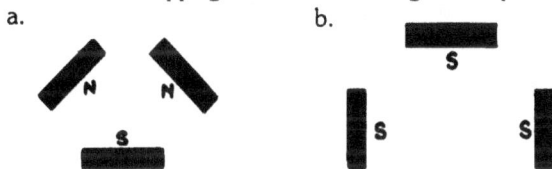

Check Point

1.

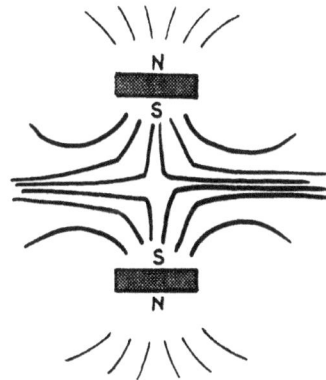

2.

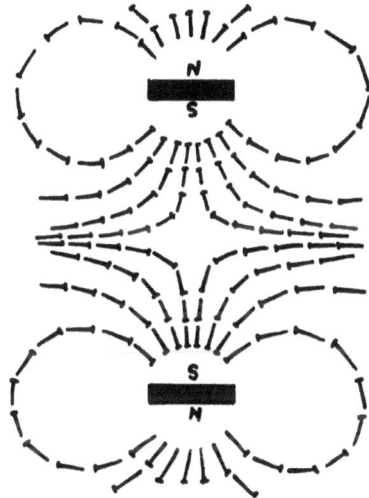

3a.

Unlike poles on the same side of the magnet should attract each other.

3b.

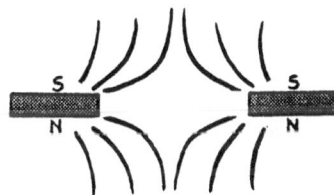

Like poles on the same side of the magnet should repel each other.

M. INVENTIONS

M-1 build an electromagnet

M-2 pin motors

M-3 dots and dashes

M-4 does it buzz it?

M-5 on-off motor

I built a real motor!

Here is an opportunity for your class to build one invention after another into a wonderful chain of understanding and achievement.

Students begin by constructing and using an electromagnet. Combining this invention with the pin compass made earlier, they build a prototype of an electric motor. This prototype rearranges to form a telegraph. The telegraph converts into a buzzer. Take the buzzer apart and build a real motor that turns with incredible speed.

All these little marvels are guaranteed to work . . ., perhaps not the first time, but with perseverance. These activities challenge students to tinker, to adjust, to refine, to trouble shoot in a logical manner until they achieve the desired click, buzz or spin. The rewards exceed the effort!

— EVALUATION —

Each question evaluates a single activity from INVENTIONS as numbered. Use any combination to frame a formal exam or an informal review: Copy these questions on your blackboard, construct your own ditto master, or photocopy the questions while masking out the rest of the page. Evaluate in ways that suit your own teaching style, enabling your students to learn and enjoy science.

Questions

M-1
Junkyard cranes use giant electromagnets to pick up old car bodies.

a. If you pulled this electromagnet apart, what would you expect to find inside?

b. As a crane operator, how would you pick up a car body and let it down again?

c. Would this crane work if you used a giant permanent magnet in place of an electromagnet? Why?

M-2
Tell how you would put together a compass, a dry cell, and an electromagnet to make the compass needle spin like a motor. Draw a diagram.

M-3
Which invention came first, the electromagnet or the telegraph? Explain your reasoning.

M-4
The hammer rings the bell by moving rapidly back and forth. What do you think the hammer is connected to? How does it work?

BELL
HAMMER

M-5
Suppose you hook your loop motor up to additional dry cells. Given enough electricity, would your motor work *without* the two permanent magnets attached to the can? Explain.

Answers

M-1
a. You would find some sort of metal core, wound with lots of insulated wire.

b. By controlling a switch that connects the wire coil to electricity: turn the switch on to pick up the car; turn it off to release the car.

c. No. If the magnet were permanent, you would not be able to turn off the magnetism to release the car.

M-2
Put the compass near the electromagnet. Touch the wires leading from this electromagnet to the dry cell as shown. To make the needle spin, turn the electromagnet momentarily on, then off, once for every revolution of the needle.

OFF
ON

M-3
The telegraph incorporates the electromagnet as one of its principle parts. The electromagnet, therefore, already existed before the telegraph was invented.

M-4
The hammer must be connected to an electromagnet. As the iron hammer moves back and forth striking the bell, it turns the electromagnet on and off.

ON: OFF:

Spring pulls up, connects electromagnet; electomagnet pulls down, breaks circuit. . .

M-5
No. Additional power would create a very strong magnetic field through the loop. But without the presence of at least one permanent magnet, the coil's powerful field would have no other field with which to interact. Being neither attracted nor repelled, the coil could not turn. (It is assumed the earth's magnetic field is too weak to matter.)

SEQUENCING

INVENTIONS follows logically from K and L. L is a prerequisite to **M** unless students first construct the compass in L-1.

Related Activities: K---L—**M**

MATERIALS

Here is everything your students will use for the next 5 activities on INVENTIONS. Materials printed in normal type are part of the core 15-things-in-a-box inventory that support all 100 activities. Materials printed in *italics* are additional local materials that you provide or ask your students to bring from home. Pencil and paper are already assumed and therefore unlisted. Each item is numbered with the activity where it is first used.

(M-1) *Medium-sized nails.*
(M-1) Plastic-insulated copper wire, about 24 gauge.
(M-1) *Scissors.*
(M-1) Size-D dry cells.
(M-1) *Masking tape.*
(M-1) *Clothespins.*
(M-2) Pin compasses constructed in activity L-1.
(M-2) Magnets.
(M-3) *Paper clips.*
(M-5) *Thread.*
(M-5) *Clear tape.*

FURTHER STUDY

Use problems like these plus "extension" ideas in INVENTIONS to lead your students beyond worksheet activity into original research and investigation. Each discovery leads to more questions, deeper questions, better questions than these. Answering them is what good science is all about.

Visit a power plant or electrical substation. Write a report about how electricity is generated in your area.

Investigate one of these career possibilities in the field of magnetism:

    physicist
    geophysicist
    surveyor
    ceramic engineer

Assume that you have overcome all technical difficulties to invent the machine of your dreams. Describe what it does. Be creative.

NAME: _____ CLASS: _____

Inventions **M-1**

BUILD AN ELECTROMAGNET

1 Get 1.5 meters of insulated wire. Measure 25 cm from one end. . .

...and begin here to wind it tightly and evenly up to the head of the nail, then back to the point.

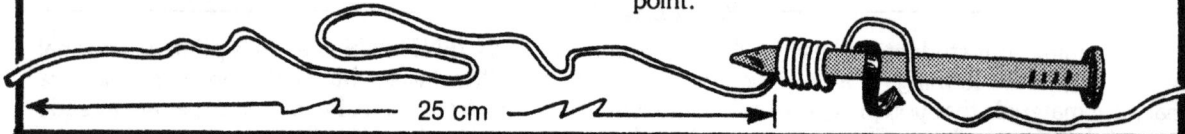

25 cm

2 Peel about 3 cm of insulation from each end. To do this, gently cut around the outside with scissors, then pull it off.

DON'T CUT THE WIRE!

SLIDE OFF

NIP GENTLY

3 Loop the longer end. . .

...then attach this loop to the flat end of a dry cell with masking tape.

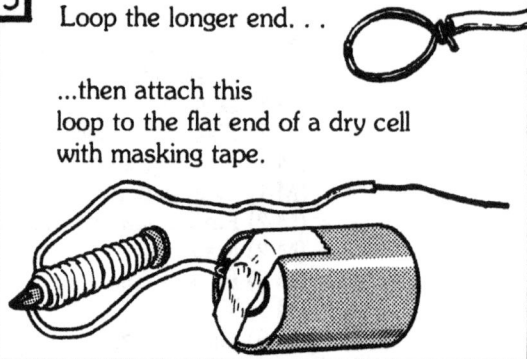

4 Put 3 clothespins on one side of your desk.

No hands?

Use only your electromagnet to pick them up and move them to the other side of your desk.

5 Tell how you picked up and released the clothespins with your electromagnet.

6 Is your electromagnet *temporary* or *permanent*? Explain.

SAVE YOUR ELECTROMAGNET

CENTIMETER RULER

cm 1 2 3 4 5 6 7 8 9 10 11 12 13 14 15 16 17 18 19 20 21 22 23 24 25

Objective

To learn how to construct and use an electromagnet. To appreciate that electromagnets are temporary, working only as electricity passes through the coil.

Preparation

Cut insulated wire into premeasured lengths about 1.5 meters (5 feet) long. If your students are working in pairs, you'll need to cut 1 length for every 2 students. Wire cutters are handy for this task, but an old pair of dull scissors will probably do. Or you can simply bend the wire back and forth until it breaks.

Peel about 3 cm of insulation from an end of a wire as directed in step 2. If the insulation is soft, you can "nip" the insulation with scissors, as instructed, or scrape it off with your fingernails. If the insulation is tough, a sharp knife may be required. Evaluate if your students can perform this task safely. If so, leave it to them. If not, strip the ends of all the wires in advance, and tell your students to skip step 2.

When you finish these activities on magnets, collect all electromagnets and save them to use again. Next time you teach these activities, you won't need to do this preparation. Take your preassembled electromagnets from storage and start right in, beginning with step 3.

Lesson Notes

1-2. An electromagnet that is prepared as directed should look similar to this:

Notice how both leads trail off the pointed end. One lead is a full 25 cm long. Both ends are free of insulation.

3. Taping one end of the wire to the dry cell makes the electromagnet easier to handle. *Never* tape both ends. This will quickly drain the dry cell of energy.

4. Unless they have had previous experience with wires and dry cells, many students will not know how to operate their electromagnets. Allow them to discover by trial and error how to turn them on and off.

Electromagnets use a lot of energy. Once students learn how to operate their electromagnets, caution them to keep them turned on no longer than necessary. Use them too much, and the dry cells won't last long enough to power other wonderful inventions still to come.

In this step, use heavy objects like clothespins for best results. Don't substitute lighter magnetic objects like paper clips. They tend to "stick" to the nail even *after* the electricity is turned off. While the nail is a strong temporary magnet, it is also a weak permanent magnet.

6. Students will use their electromagnets again when building motors, telegraphs and buzzers. They should store them completely separated from their dry cells to avoid accidentally draining their cells of energy.

Extension

When electrons move they always create an associated magnetic field. Electrons that spiral around a coil, for example, produce a magnetic field that is directed through the length of this coil. While this field is strengthened by an iron core, (a nail, for example) its presence is not necessary.

Electrons that travel around a wire loop, for example, also create a magnetic field directed through the loop. You can detect the presence of this field with a compass needle. Such a device is known as a galvanometer (electricity tester).

To build a galvanometer, tape a loop of insulated wire to the cover of a book. Hang a magnetized pin from a short strand of human hair (not thread), and tape it to the top of the loop. Turn the book so the pin hangs parallel to the plane of the loop.

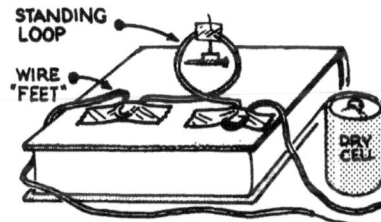

Now connect the loop momentarily to a dry cell. How does your galvanometer detect electricity? (Electrons moving around the coil create a magnetic field that is directed through the coil. The needle deflects to line up with this field.) How can you make the needle jump the other way? (Reverse the cell so electrons travel in the opposite direction.)

Check Point

5. Complete the circuit by touching the wire leads from the electromagnet to a dry cell. This creates a temporary magnetic field through the nail, allowing you to pick up a clothespin by its iron spring. To release the clothespin, simply break the circuit. This stops the flow of electricity, breaking the associated magnetic field. The clothespin drops immediately.

6. It is a temporary magnet. The field is only maintained as electricity flows through the coil. Break the circuit and you turn off the magnetism.

PIN MOTORS

1 Set up the pin compass you made in activity L-1.

2 Tape a magnet to the end of your pencil like this.

3 Turn the magnet back and forth to spin your compass.

4 Explain how to keep the compass spinning.

5 Put a second clothespin on your can, a quarter turn from the compass.

¼

6 Clamp your electromagnet from the last activity in this second clothespin. Adjust so the compass pins turn near the head of the electromagnet.

7 Make your "motor" spin by turning the electromagnet on and off. Explain how it works.

SAVE YOUR MOTOR.

Objective

To build two simplified models of an electric motor. To understand how they work.

Introduction

None required.

Lesson Notes

3. For each revolution of the pins, it is necessary to turn the magnet back and forth through one-half turn. Changing the poles in this manner attracts, then repels, each pin as it turns past the magnet. Keep changing the poles to keep the "pin motor" spinning.

Some practice may be necessary to develop a sense of timing. Choose a spot on the compass, the"N" for example. Keep your eyes glued to this spot, flipping the magnet back and forth once every time the "N" turns through one complete revolution.

6. Keep the gap between the rotating pins and the electromagnet about as wide as your pencil.

7. As in step 3, the pins spin because of variations in the applied magnetic field. Instead of shifting the field mechanically, here the field is turned on and off.

Check Point

It's difficult for students to explain how they keep the pins spinning. Like riding a bicycle, it's easier to do than to explain. Encourage students to verbalize their answers first, then write down what they've said.

4. The magnet attracts one of the pins and repels the other. The attracted pin turns toward the magnet as near as it can get while the repelled pin turns away as far as possible. Just *before* this happens however (and the pins stop), you reverse the poles by turning the magnet. The closest pin is now repelled on around while the far pin is now attracted. So the pins continue to spin around another half turn to where you reverse the poles again. By repeating this process over and over, the pins keep rotating.

7. Touch the wires to the dry cell for just an instant. This creates a magnetic field, attracting one of the pins and repelling the other. The pins spin one full revolution to where you can turn the electromagnet on to attract (and repel) them again.

DOTS AND DASHES

1 Remove the compass from your pin motor. Slide the clothespins to opposite sides.

2 Strip the insulation from 20 cm of wire.

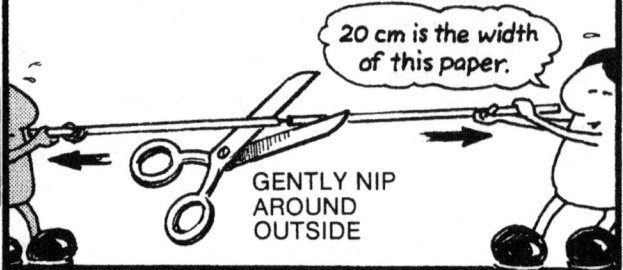

20 cm is the width of this paper.

GENTLY NIP AROUND OUTSIDE

3 Wrap this bare wire about 5 times around a paper clip. Clamp the other end in the unused clothespin.

4 Adjust this paper clip so it is free to hit against the head of the electromagnet.

VERY SMALL GAP

5 Make your telegraph click on and off.

TROUBLE-SHOOTER CHECKLIST...

6 Have someone ask you questions. Answer them on your telegraph.

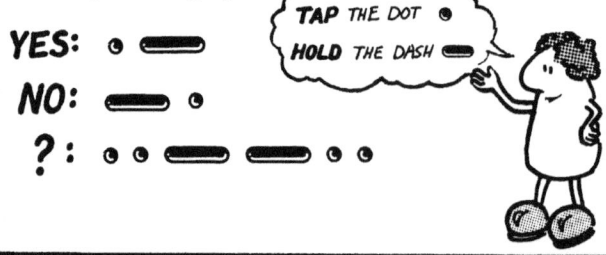

YES: • ▬

NO: ▬ •

?: • • ▬ ▬ • •

TAP THE DOT •
HOLD THE DASH ▬

If the paper clip *won't move:*

➤ Make the gap smaller.

➤ Lengthen the paper clip wire.

➤ Turn the dry cell around.

➤ Use 2 cells in series.

If the paper clip *sticks* to the nail:

➤ Be sure the nail touches only copper wire, not the iron paper clip.

➤ Wrap the copper wire 1 or 2 more turns around the paper clip.

7 Tell how your telegraph works.

SAVE your telegraph parts to make a buzzer!

Objective

To build a working model of a telegraph. To understand how it works.

Introduction

Build this telegraph yourself before your students try it. This will familiarize you with the directions and provide a model for your students to follow.

Lesson Notes

The next three activities all involve inventions: a telegraph, a buzzer, and a motor. Your students have the opportunity to become expert engineers in a world of paper clip technology — to twist, turn, and adjust until the telegraph clicks, the buzzer vibrates, and the motor turns.

Each activity requires a certain amount of eye-hand coordination. Be prepared for lots of initial excitement, followed by some it-won't-work frustration, followed by I-finally-did-it triumphs.

Activity M-3 is the easiest; activity M-4 a little harder; activity M-5 the most difficult. If your class experiences too much difficulty building this telegraph, consider studying the other inventions in a different context, perhaps as a teacher demonstration, or as a science project for selected students.

2. This cut-and-pull technique may not work with your particular brand of wire. If your insulation is tough enough to require stripping with a knife, you may wish to complete this step yourself. If bare copper wire is available, use that instead and skip this step entirely.

3. The paper clip and electromagnet tend to become permanently magnetized when touched together. This makes them stick together even after the telegraph key is released. To overcome this problem, the paper clip is first wrapped in copper wire. When the telegraph key is depressed, non-magnetic copper strikes the nail head, not the iron clip. The paper clip remains free to spring back when the telegraph key is released.

4. The paper clip and wire should meet the electromagnet at an approximate right angle. This allows it to swing freely toward the electromagnet and spring away.

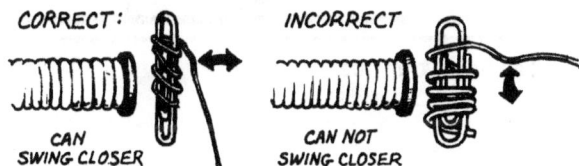

CORRECT: CAN SWING CLOSER   INCORRECT: CAN NOT SWING CLOSER

Keep the space between the paper clip and nail head as narrow as possible. Reposition the electromagnet, bend the wire arm — do whatever is necessary to make them almost touch.

5. The wire arm acts like a spring, pulling the paper clip away from the electromagnet when you release the telegraph key. If this wire is too long, its spring tension may be too weak: the clip will stick to the electromagnet or double-click against the nail head before it springs away. If this wire is too short, its spring tension may be too strong: the clip may not move far enough forward to click against the electromagnet at all. Making trial and error adjustments, your students must find just the right tension, shortening or lengthening the wire until the telegraph clicks just right.

Some students, at the first sign of trouble, throw up their hands and run to the teacher for help. Discourage this. A trouble-shooting table is provided in this activity to help students solve their own problems. Those who persevere and succeed will gain self-confidence in their own problem solving ability.

6. Tap out these yes-no-maybe responses on your desk with a pencil for all to hear and identify. Then ask your students questions and have them tap back yes, no or maybe in code response.

Extension

For those who wish to telegraph more extensive messages, provide copies of this international code.

A .—	N —.	1 .————
B —...	O ———	2 ..———
C —.—.	P .——.	3 ...——
D —..	Q ——.—	4 ....—
E .	R .—.	5 .....
F ..—.	S ...	6 —....
G ——.	T —	7 ——...
H	U ..—	8 ———..
I ..	V ...—	9 ————.
J .———	W .——	0 —————
K —.—	X —..—	(.) .—.—.—
L .—..	Y —.——	(,) ——..——
M ——	Z ——..	(?) ..——..

ABBREVIATION:	MEANING:
V	From
AA	Who are you?
K	End of message; answer
R	Message received and understood
N	Negative — No
A	Affirmative — Yes
IMI	Repeat
Series of dots	Error, will resend word

Check Point

7. Turning the electromagnet on (completing the circuit) creates a temporary magnetic field in the nail. This field attracts the iron paper clip, causing it to click against the nail. Releasing the wire breaks the circuit, stops the flow of electricity and thereby breaks the magnetic field. No longer attracted, the paper clip springs back off the nail.

DOES IT BUZZ IT?

1 Pull apart your telegraph. Clothespin the longest lead from your electromagnet over the lip of the can like this.

LONGER LEAD

CLIP OVER EDGE OF CAN.

2 Push up the long wire so the electromagnet swings free and level. Point the short wire straight back.

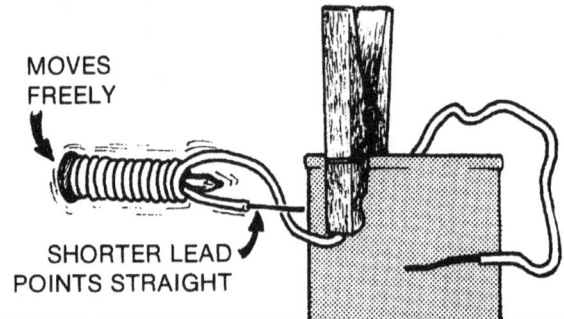

MOVES FREELY

SHORTER LEAD POINTS STRAIGHT

3 Clothespin the other copper wire (with its paper clip) over the lip of the can like the first. Place it so the clip can swing against the head of the electromagnet.

Leave a TINY GAP.

TROUBLE-SHOOTER CHECKLIST!

If the paper clip *won't move*:

- Make the gap smaller.
- Lengthen the paper clip wire.
- Turn the dry cell around.
- Use 2 cells in series.

If the paper clip *sticks* to the nail:

- Be sure the nail touches only copper wire, not the iron paper clip.
- Wrap the copper wire 1 or 2 more turns around the paper clip.

4 Buzz your buzzer. Hold the long wire (A) to the flat end of a dry cell while you press the bump end *gently* against the short wire (B).

Bz–z–z

B

A

Touch VERY LIGHTLY

5 Tell how your buzzer works.

Objective

To build a working model of a buzzer. To understand how it works.

Introduction

Build this buzzer yourself before your students try it. This will familiarize you with the directions and provide a model for your students to follow.

Lesson Notes

1. Notice that the wire bends over both the inside *and* outside rim of the can, so it can be firmly clamped by the clothespin.

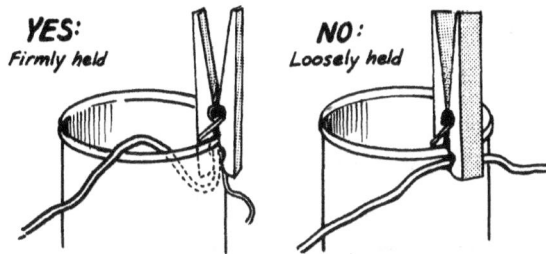

YES:
Firmly held

NO:
Loosely held

3. As with the telegraph, the paper clip must meet the electromagnet at an approximate right angle. This allows it to swing freely towards the electromagnet and spring away. This back and forth motion is rapid enough to create a vibrating buzz.

4. The nail almost touches the copper wire wrapped around the paper clip. Students will need to make fine adjustments —turn the electromagnet or bend the wire arm — to bring them very close together.

Some may hold onto the free-swinging electromagnet. If they do this, the buzzer won't sound because the electromagnet can't vibrate. Holding the electromagnet converts the buzzer, in effect, into a telegraph.

To operate the buzzer, hold *only* the cell and press it *gently* against the electromagnet. This moves the electromagnet nearer to the paper clip, causing both to vibrate. The movement of the electromagnet against the dry cell creates the on-again-off-again magnetic field that sustains the vibration.

Again, a trouble-shooting chart is provided so that students can experience the satisfaction of solving their own problems.

Check Point

5. As the paper clip and electromagnet separate, lead "B" bumps into the dry cell and thereby completes the circuit. This creates a magnetic field that attracts the paper clip and electromagnet back together. As they click together, lead "B" gets pulled off the dry cell thereby breaking the circuit. This turns the magnetic field back off, allowing the paper clip and electromagnet to spring back apart. This, in turn, closes the circuit and the cycle repeats. This back and forth response to the on-again-off-again magnetic field is so rapid, it creates a buzzing sound.

NAME: CLASS:

ON-OFF MOTOR

1 Straighten 2 paper clips.

Use them to put *attracting* magnets across the rim of a can like this:

tape
rubber band
N
S

2 Use this ruler to measure out 2 pieces of wire, each 24 cm long.

Peel 4 cm of insulation from all 4 ends.

Wrap an end of each wire around your pencil.

2 LOOPS

3 Place these loops above the can, level with both magnets.

Put under the rubber band, and add tape below.

4 Wrap another 24 cm piece of wire around a dry cell.

Remove the coil, but don't let it get bigger.

5 Tightly loop each end like this:

⅔ SINGLE WIRE
TWIST
⅓ DOUBLE WIRE

Peel off the insulation from each arm.

6 Tie a thread to the loop. Wind it 2 times around the arm, and tape the end. Trim with scissors.

TRIM   TRIM   TRIM

Do one side only.

7 Shape this coil so it turns easily on the loops between both magnets.

8 Connect the 2 long wires to a dry cell. Blow on the coil to start your motor spinning.

If it won't work, try reversing the coil.

9 What makes the coil spin? (Hint: why put thread on one of the arms?)

CENTIMETER RULER

Objective

To build an on-off motor that spins under its own power.

Introduction

Build this on-off motor yourself before your students try it. This will familiarize you with the directions and provide a model for your students to follow.

Lesson Notes

2,4. Gently cut around the insulation with a pair of scissors, then pull it off. If the insulation is too tough to do this, strip it off with a sharp knife. You may need to do this yourself depending on the maturity of your students.

4-5. Use the correct wire length —24 cm (9 1/2 inches) — no more and no less. When you wrap it 1 1/3 times around a size-D dry cell, it forms a coil with just the right dimensions.

This coil is held in place by looping two wire arms to the outside, as shown. Notice that 1/2 of the coil contains a double strand, while 2/3 of the coil has only a single strand. This distributes the weight evenly above and below the arms, insuring that the coil will spin smoothly and easily in step 7.

6. Most electric motors have commutators that change the direction of the magnetic field back and forth. In place of a commutator, this motor substitutes a single thread wrapped around *one* arm of the coil. Every time the coil turns, it bumps over the thread and momentarily breaks the circuit. This creates an on-off variation in the magnetic field of the coil that keeps it spinning.

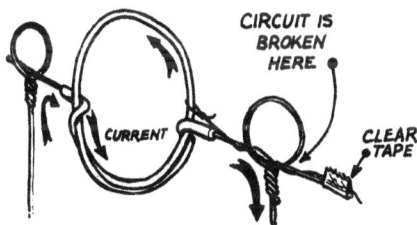

Trim back the tape that holds the thread until there is only a small piece left. Clear tape is easier to use in this step than masking tape.

7. It is crucial that the coil spins smoothly and easily. If it is out of balance, reform the coil to more equally distribute the weight of the wire above and below the support arm. Younger students may require extra assistance here.

Raise the coil high enough so that an imaginary line from the middle of one magnet to the middle of the other passes through its center.

Keep the magnets as close as you can without touching the spinning coil. Be sure these magnets *attract* with *opposite* poles facing each other. The motor will not work if the magnets are set in a repelling position. (This is a common error.)

8. When you connect the motor to the cell, the loop may give a little kick or jerk. This is a good sign: electricity is flowing through the coil, creating an associated magnetic field that is interacting with the two permanent magnets.

Now spin the coil by twisting one of its arms with your fingers or by blowing on it. If you're lucky, the coil will keep on spinning. If not, shift the coil a little right or left so its arms contact the support loops in a slightly different position. Then try again.

You will eventually find a place where the string breaks the magnetic field in just the right place. This allows the coil to spin *beyond* its point of greatest attraction while turned off, only to turn back on to be attracted around again.

Over extended periods of time, bare copper wire tends to oxidize, insulating its surface to the passage of electricity. If the arms of the loop have not been recently stripped, or if they appear a dull cloudy-green color, scrap away this oxide covering with a knife or nail to expose shiny new metal underneath. This may fix coils that refuses to turn, or improve the performance of motors that turn slowly.

If a motor proves stubborn beyond all remedy, try zapping it with two cells in series. Or if it turns especially well with only one cell, see if it will run on only one magnet as well. If the coil is well balanced, it should still spin.

Check Point

9. The flow of electricity creates a magnetic field that turns the coil toward an attracting magnet (away from a repelling magnet). Then it crosses the thread. This *momentarily* shuts the field down. The coil spins on by, only to turn itself back on and be attracted (and repelled) again.

N. ANIMALS

N-1 bug in a jar

N-2 looking at animals

N-3 adapt-a-bird

N-4 camouflage

N-5 to run or not to run

N-6 beat it!

These activities invite students to observe common animals in their environment, to compare and contrast what they see, to draw conclusions. Special emphasis is placed on adaptations — how the animal's physical structure or behavior enhances its ability to survive. Does it hop, crawl, swim or fly? Will it claw and chew its way through life or hide and blend in? Does it carry heavy armor, sport quills, sting or just smell bad?

Your class will cut and paste birds into specific environments where they can survive. They'll camouflage moths so perfectly that not even big bird (the school headmaster) can find them. They'll design animals with special adaptations to get them out of all manner of tight spots.

These activities speak to the creativity inside each of us. They form a special blend of art, science and imagination.

EVALUATION

Each question evaluates a single activity from ANIMALS as numbered. Use any combination to frame a formal exam or an informal review: Copy these questions on your blackboard, construct your own ditto master, or photocopy the questions while masking out the rest of the page. Evaluate in ways that suit your own teaching style, enabling your students to learn and enjoy science.

Questions

N-1
Compare and contrast your hand with your foot.

N-2
Look at the clock in your room or a wristwatch.
a. Write an observation about this clock.
b. Write an hypothesis about this clock.

N-3
Where does this bird likely live, and how does it eat? Explain.

N-4
In England, during the industrial revolution, factories burned so much smoky coal that the countryside gradually became darker over a long period of time! How do you think the moths in this area responded to their slowly darkening environment?

N-5
How can BOTH of these statements be true?

a. Cheetahs can outrun humans.
b. Humans can outrun cheetahs.

N-6
You are an animal that lives among sharks in the ocean. Draw a body that helps you escape being eaten. Write about how you survive, using complete sentences.

Answers

N-1
Accept any thoughtful answer that contains both a comparison and a contrast. Example: Both the hand and the foot have five digits each. The hand, however, has an opposable thumb that is suitable for grasping, while the foot does not.

N-2
a. The second hand sweeps continuously around the clock face.
b. The second hand is driven by an electric motor linked to gears.

N-3
The bird's long legs are adapted to standing in shallow water, and its webbed feet for swimming. It probably uses its long neck and shovel-like beak for scooping up plants and animals that live on the bottom of shallow ponds.

N-4
The moths gradually got darker, too, better blending in with their changing environment. (More capable students might further explain how this process works: Lighter colored moths were naturally selected by birds over darker colored moths, because they were easier to see. A greater proportion of darker-colored moths survived to pass their more adaptive coloration on to new generations of moths.)

N-5
a. Cheetahs have greater speed. They can run faster than anything on legs *if* the race covers only a short distance.
b. Humans have greater endurance, the ability to run for many hours at a moderate, steady speed. They can outdistance cheetahs *if* the race is long enough.

N-6
(Accept any answer that thoughtfully explains how body structure helps the creature survive.)

SEQUENCING

ANIMALS may be studied at any time. It has no prerequisites.

Related Activities: **N**

MATERIALS

Here is everything your students will use for the next 6 activities on ANIMALS. Materials printed in normal type are part of the core 15-things-in-a-box inventory that support all 100 activities. Materials printed in *italics* are additional local materials that you provide or ask your students to bring from home. Pencil and paper are already assumed and therefore unlisted. Each item is numbered with the activity where it is first used.

(N-1) *Bugs in jars*.
(N-1) *String* (optional).
(N-1) *Animals* to observe.
(N-1) Scissors.
(N-3) Clear tape.
(N-4) *A coloring system*: colored pencils, crayons, tempra paint, or watercolor sets.
(N-5) A *wall clock* or watch with second-hand sweep.

FURTHER STUDY

Use problems like these plus "extension" ideas in ANIMALS to lead your students beyond worksheet activity into original research and investigation. Each discovery leads to more questions, deeper questions, better questions than these. Answering them is what good science is all about.

Find out all you can about one of these creatures. Report on its special survival strategies:

octopus	mongoose
squid	kangaroo
giraffe	hippopotamus
elephant	flatfish
opossum	cuttlefish
lady bug	chameleon

Brainstorm a list of at least 20 inventions and discoveries that have both helped and harmed the ability of humans to survive on this planet. Then order these technologies, ranking positives at the top and negatives at the bottom.

Find out what a naturalist does and write a job description. How would you like to have such a job?

NAME: _____   CLASS: _____

BUG IN A JAR

1 Compare yourself to a bug in a jar.

Identify yourself: _____

Identify your bug: _____

COMPARE: List 5 ways you are the **same** as your bug:

a. _____

b. _____

c. _____

d. _____

e. _____

CONTRAST: List 5 ways you are **different** than your bug:

f. _____

g. _____

h. _____

i. _____

j. _____

2 Measure yourself and your bug with this ruler. You may want to use some string as well.

a. Your Height: ....

b. Arm Length: ....

c. Explain how you measured yourself:

a. Bug's Length: ....

b. Front Limb Length: ....

c. Explain how you measured your bug:

3 How many bugs lined up head to tail would equal your height?

Show your math.

4 If your bug were as tall as you, would its arms be longer or shorter than yours?

Show your math.

cm ruler

Objective

To observe similarities and differences between yourself and another animal.

Preparation

Ask each student to bring a live bug to school in a small, clean jar. Students who forget to do this might pair up with others who remember. If your room is a small animal zoo already, perhaps you can spare outside creatures altogether and use your own resources.

Lesson Notes

Remind students to treat the bugs they catch with due consideration and respect. (If someone put you in a jar, how would you like to be treated?) They should handle them gently, of course, and keep them in captivity no longer than necessary. If students need to keep their bugs for more than a few hours, they must ventilate the jar with air holes in the lid, and perhaps add a bit of soil or vegetable matter to provide moisture and food.

2. The ruler, of course, is not long enough to measure human dimensions. String can help. Students might cut lengths equal to their height and arm length, then count by 20s — 20 cm, 40 cm, 60 cm, . . . until they reach the end of the string.

Younger students may lose track as they count. Use this conversion table to detect gross measuring errors:

$$4.0 \text{ feet} = 122 \text{ cm}$$
$$4.5 \text{ feet} = 137 \text{ cm}$$
$$5.0 \text{ feet} = 152 \text{ cm}$$
$$5.5 \text{ feet} = 168 \text{ cm}$$
$$6.0 \text{ feet} = 183 \text{ cm.}$$

It's not easy to measure the length of a bug's arm, but students can estimate. They might draw lines on paper, for example, that match the dimensions of their particular bug, then measure the lines they have drawn. Or they could draw a grid and observe their bug walk across it. Give students the space and freedom to come up with their own solutions.

3-4. These last two steps require students to make quantitative sense of the measurements they have just made — to compare and contrast body size in a meaningful way.

Let your class first try these problems unassisted. If they need help, write these more detailed instructions on your blackboard:

(3) Divide your bug's length into your own length to find a ratio (how many times it fits).

(4) Multiply this ratio by the bug's limb length. Is the result longer or shorter than your own arm?

As a last resort, work out the problems as a class exercise. Your task as the teacher is to provide the minimum amount of help necessary, to enable your students to grow intellectually and experience success.

If some animals look sluggish or sick, ask students to release them in a suitable place on the school grounds, or to take them home and put them back under the rock or log where they were found. If they are active and healthy, consider saving them for the next activity.

Extension

Write an account of where you looked for your bug and how you captured it. What kinds of escape strategies did the bug use to try to get away from you?

Check Point

1. There are numerous similarities and differences to notice. Size, color, shape, texture, structure, body parts, function, and movement are only some of the possible categories to compare and contrast. Redirect those who limit their observations to single categories, color, for example: I have brown eyes, it has red eyes; I have black arms, it has brown arms; I have white fingernails . . ., etc. Encourage higher, more complex levels of thought and expression whenever possible.

2. Answers in steps 3 and 4 are based on these particular bug measurements:

$$\text{student's height} = 150 \text{ cm}$$
$$\text{arm length} = 60 \text{ cm}$$
$$\text{bug's height} = 1 \text{ cm}$$
$$\text{front limb length} = .5 \text{ cm}$$

3. student's height/bug's height = 150 cm/1 cm = 150

4. 150 x .5 cm = 75 cm>60 cm. Thus, proportional to body size, the bug's arm is longer.

NAME: _____   CLASS: _____

LOOKING AT ANIMALS

CUTOUTS

N-2

| **Name of animal:** | **Place observed:** | **Date observed:** |

cm ruler

1 Describe your animal. Use complete sentences.

a. Color:

b. Size:

c. Shape:

d. Your feelings about this animal:

2 Draw your animal here. Sketch details accurately.

3 **OBSERVATION:** Take 5 minutes to record everything you notice about how your animal moves and sounds. Write **WHAT** you see or hear, but don't explain it.

(*example:* **1.** It walks in circles.)

An OBSERVATION describes WHAT happens.

1. ....
2. ....
3. ....
4. ....
5. ....
6. ....
7. ....
8. ....
9. ....
10. ....

4 **HYPOTHESIS:** Try to explain **WHY** the animal moves or sounds like it does. Choose numbers from above, then write your reason.   (*example:* **1.** It was trying to escape.)

An HYPOTHESIS explains WHY something happens.

Objective

To provide a generalized form useful for observing a diversity of animal life, both in and out of the classroom. To understand the distinction between an observation and an hypothesis.

Preparation

Use bugs from the last activity if they are still active, or find new specimens. Students might also observe larger animals — goldfish in an aquarium, a pet rabbit, etc. Any subject that moves or makes noise is suitable. You can even ask the class to observe you, the teacher, observing them. (A nice way to enjoy 5 minutes of silence!)

Lesson Notes

This activity invites your students to observe animals on many levels: to describe and draw them; to concentrate for five uninterrupted minutes on their movements and sounds; to record events as they actually happen; to make sense of the behaviors they see.

2. Creativity should be consistently attempted at many levels, graphic as well as verbal. If you honor only the spoken word or the written idea, you limit other legitimate ways of knowing. Everyone can draw. It only takes time and a little serious effort. You must look clearly, process what you see, then output your knowledge in a calm, accurate manner.

Demand as much as students are able to give. Even young students can produce drawings that pay some attention to detail. Encourage older students to specify scale and point of view — top view, side view, etc.

3-4. Writing down each thing you notice in 5 minutes of single-minded concentration may seem difficult at first, but it's well worth the effort and practice. As students learn to concentrate and record data, this research skill can begin to eliminate the ancient game of copying from reference books and calling it science.

Here is an interesting way to distinguish between observations and hypotheses. While you have the undivided attention of your entire class, write this sentence on the board: "Why am I writing this?" After a few moments of mysterious silence, erase it.

Ask your students to tell you what you did, sorting out personal speculations from specific observables. Those who are *hypothesizing* will attempt to answer the question. Those who are *observing* will not answer the question, but rather describe you writing it.

Extension

Don't limit animal observations to your classroom. This worksheet is also available as a cutout so students can use it in other settings, observing animals wherever they are found. Here are some suggestions.

a. Ask students to record multiple animals observations outside of school over an extended period of time, modeling each new report to the format and structure of this worksheet. Multiple observations could be combined to form part of an ongoing journal.

b. Organize a field trip. Any natural setting will provide a rich variety of things to see and hear, to wonder about and explain.

c. Assign home observations of family pets.

d. Challenge students to observe the same animal, in the same location, at the same time each day. Birds are creatures of habit.

The possibilities for growth are rich indeed, both in science and in language arts.

Check Point

Review sentences and drawings for attention to detail and completeness of thought. Check that students have distinguished between "what" observations and "why" hypotheses.

ADAPT-A-BIRD

CUTOUTS
N-3
2 pages

Cut and tape a bird into each habitat.

Cut out body parts that will help each bird survive in its environment.

Explain why you choose certain body parts.

WATER ENVIRONMENT		

DENSE FOREST 		
OPEN SPACES 		

Objective

To appreciate how each body part helps a bird survive in its own particular environment.

Introduction

Ask your class to imagine for a moment that they are all birds. On a piece of scratch paper, have them write down their food of preference, then draw the best possible beak shape to help them easily obtain this food — long and slender? blunt and short? hooked? curved? scooped? Emphasize that details are important. The shape of the bird's beak may determine whether it lives or dies. Ask volunteers to share their answers with the class.

Lesson Notes

Both this activity sheet and the supplementary sheet of bird body parts that supports it are available as student cutouts. You can also reproduce the body-parts page, if you so choose, from the line master found at the back of this book.

Students should cut out and assemble a complete bird into each of these three environments, clear-taping body parts directly onto each picture (not below it). These pieces may also be glued, but the individual body pieces will be harder to position with any great accuracy.

Discourage the practice of cutting out all the bird parts first. This is time consuming and unnecessary. Not all parts will be used. Moreover, even a slight breeze could cause great disorder and carry away many of the smaller pieces.

Precision cutting is not necessary. Even rough cuts work well, outlining the bird in a jagged white background. Some of the birds may look ill-conceived, but don't worry. More important, does the explanation under each picture justify the selection of body parts in a thoughtful, coherent way? If so, the bird survives!

Check Point

Here is one possible set of answers. There are, of course, many possible adaptations:

This bird uses its large wingspan to fly high and easy over open spaces, looking for small animals. Its feet have claws for clutching prey, and its beak is curved to tear up meat.

OPEN SPACES

This bird is small enough to easily dart in and out of tight spaces among trees. Its body has a camouflaged pattern to blend in. Its feet can grasp small branches. Its short, stout beak enables it to crack open seeds and search out insects.

DENSE FOREST

This bird stands in shallow water on its long legs. Its webbed feet keep it from sinking into the mud and also help it swim. Its long neck enables it to feed off the bottom, scooping up water plants into its shovel-like beak.

WATER ENVIRONMENT

CAMOUFLAGE

1 Cut out a moth square.

Leave the moth in its square.

2 Choose a resting place for this moth that you can easily see from the center of the room.

The bulletin board is easy to see...

Carefully paint your moth to look exactly like its resting place.

Great camouflage!

3 Cut out your moth and write your name on the back.

4 Fix your moth to its resting place with a small piece of tape rolled sticky-side out.

5 Keep track of time. Moths must be in place before the birds come!

6 Which kind of animal needs to use camouflage: the hunter, the hunted, or both? Explain.

7 List advantages and problems of using camouflage.

ADVANTAGES	PROBLEMS

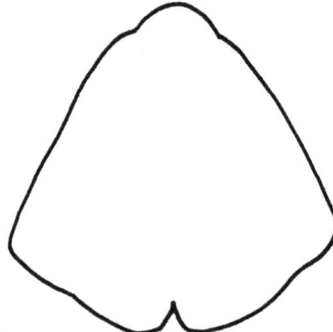

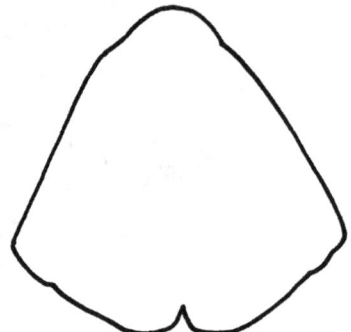

Objective

To camouflage paper moths so they blend into the patterns and textures of classroom surfaces.

Preparation

Gather together materials for coloring and camouflaging moths. Use colored pencil, crayons, tempra paint, watercolor sets, whatever is available. If you decide to use paint, provide blue, yellow and red primaries plus black and white. Students can mix and match any color they need from these basics.

Lesson Notes

Careful, thoughtful work requires considerable time. Students may need a whole period to create just *one* well-camouflaged moth (steps 1-5), to be visited by the "birds" (between steps 5 and 6) and to debrief (steps 6-7). If this is all the time you can afford to spend, it is better to limit your students to one moth done well than three moths done poorly.

Students can, of course, make several attempts, then expose their best moth to the scrutiny of the birds. If possible, spend an entire class period camouflaging all three moths well, then arrange to have the birds come on another day.

1-4. Notice that the entire square is first cut out (step 1) with the moth shape still inside. It is then fully camouflaged (step 2) *before* being trimmed (step 3) and set in place (step 4). In this manner, desks and walls are protected from any color that might overlap outside of the moth outline. This overlap is then easily cut away. Remind students to trim just *inside* the black line. A well-camouflaged moth doesn't show its outline.

2. Students should pick moth-resting sites with great care. The survival of each moth depends on it. Stress these ground rules: Moths must be plainly visible from the center of the room, not resting under or behind things, or otherwise out of view. They must be within easy reach, not taped to a 10-foot high ceiling or fixed to the outside of a window.

To match colors and textures, students need the freedom to stand close to the surface they wish to imitate. Encourage them to test color blends on pieces of scratch paper *before* painting their moths.

5. "The birds are coming. The birds are coming. Get ready for the birds!" As the time approaches, you can generate great anticipation and excitement. Students may question if the birds are real. Be vague and cagey about your answer. "Of course they are real, alive and breathing, all of them!"

Students need to manage their time well. Let them know when the birds will arrive. Stay on schedule. If the moths aren't ready, too bad . . .

Here is a wonderful opportunity to introduce other faculty, even the school principal, to the wonders of hands-on science. Prearrange to have a group of 3 or 4 students and/or adults "scratch" on your door at the appointed hour. Show them a moth pattern in advance, so they know what they're looking for. Tell them that the moths will be in plain view about your room, but difficult to see.

When you open the door, the birds should swoop into your room and gobble up as many moths as they can find before you shout "TIME." Each time they find one, they should remove it from its resting place and shout "DINNER" or some such phrase.

Determine how many moths are in place before the birds arrive. Count the number of times you hear the word "DINNER" during the hunt, so you can call "TIME" before all the moths are eaten. Allow at least 10% of the moths to survive, perhaps more if they are well-camouflaged.

6-7. Shoo away the birds as quickly as they came. Make sure they leave all "prey" on a designated desk top. Discuss why some moths survived and others did not. Are luck and location important factors? Discuss concepts of adaptive coloration and camouflage, reserving enough time for students to thoughtfully answer each question. You may wish to honor the survivors in a special bulletin board display.

Check Point

6. Both use camouflage. The hunting lion, owl or spider needs to surprise its prey; the hunted antelope, field mouse or moth needs to escape being noticed.

7. Advantages: Camouflage is easy to use. Just stay still. More elaborate forms of protection like speed, protective covering or poison may not be necessary.

Problems: Camouflage doesn't work as well when moving. Nor will it work in other environments, against a snowy background, for example.

NAME: CLASS:

TO RUN OR NOT TO RUN

1 Do this experiment with a friend. Decide who will be the cheetah and who will be the zebra.

ME CHEETAH   ME ZEBRA

2 The **cheetah** must run quietly in place as **hard** as possible. The zebra times how many seconds pass before the cheetah slows down.

a. Time before cheetah slows:

seconds

b. As a cheetah you travel 30 meters each second. How far did you go?

3 Now the **zebra** must jog at an **easy** pace. The cheetah times how many seconds pass before the zebra slows down.

a. Time before zebra slows:

seconds

b. As a zebra you travel 15 meters each second. How far did you go?

4 The cheetah (fastest land animal on earth!) catches something to eat only once in 7 tries.

Use your results to explain why.

5 You are a rabbit that survives by using camouflage.

Have a friend time how long you can remain **perfectly still.** Get comfortable and signal when to start:

Timing stops at the **first movement** of any kind. Blinking and breathing are OK, but not smiling, shifting, scratching, etc.

a. My time:

trial 1	trial 2

b. Friend's time:

trial 1	trial 2

Find a **CLASS CHAMPION:**

6 Imagine you are a cheetah, a zebra or a rabbit. Write a story about who you are and how you survive.

Copyright © 1988 by TOPS Learning Systems. Reproduction limited to personal classroom use only.

Objective

To examine speed, endurance, and camouflage as survival techniques. To consider the trade-offs.

Introduction

This activity is full of action, great fun. Your students will run flat out like a cheetah, gallop along like a zebra, or sit perfectly still like a rabbit. To control youthful exuberance, you'll need to assume a serious, no-nonsense demeanor.

Before students begin, emphasize the difference in animal running styles. Cheetahs run hard, flat out, holding nothing back. (Ask a volunteer to demonstrate by running furiously in place for perhaps 5 seconds.) Zebras, however, run slow and steady. (Ask the same volunteer to run easily in place for another 5 seconds.)

Lesson Notes

1-4. If students adopt the correct running style, "cheetahs" will slow down dramatically and show obvious signs of fatigue within one minute. "Zebras", however, should be able to run easily for the full five minutes without getting tired.

If students don't adopt the correct running style, it is possible that a cheetah will occasionally go further than the zebra before getting tired. (At twice the speed, it only needs to run half as long). That's okay. The cheetah, after all, catches something every now and then (about once in seven tries).

All running times are expressed as seconds, not minutes. Students typically forget to make this conversion, especially when measuring the longer zebra running time.

Be aware that this analysis, based on distance, only works to the zebra's favor if it has a good head start. At close range, the zebra doesn't have time to make the cheetah tired. Speed wins out over endurance if the cheetah can sneak up close enough before attacking. Thus, cheetahs work at being very sneaky.

5. Remaining perfectly still is not easy, especially when you're only a laugh or smile away from having time called on you. That's why each student has two chances or trials. Champion rabbit honors will likely go to the best poker face.

Extension

Copy this story starter on your blackboard for the class to finish. Assign this creative writing exercise only *after* your class has completed all 6 animal activities in this series.

MEANWHILE, BACK IN THE JUNGLE . . .

I had come to realize that I must leave the cave. It had sheltered me for the past weeks while my leg was on the mend. But my food and water were now gone.

Between me and the safety of my mountain village were many obstacles. Hungry lions were present in the tall savannah grass, the river beyond was full of crocodiles, the distant forest contained unknown dangers.

I limped out into the bright, warm sunlight . . .

Students should incorporate as many animal survival strategies into their writing as possible. The goal is to create excitement and complexity that will maintain high reading interest. Good science and good communication go together.

Check Point

2-3. Here is one possible result for the cheetah slowing after 60 seconds, the zebra running the full 5 minutes:

cheetah: 60 sec x 30m/sec = 1,800 m

zebra: 5 min x 60 sec/min x 15 m/sec = 4,500 m

4. Even though the zebra runs slower than the fastest land animal on Earth, it can still usually escape. The zebra expends much less energy than the cheetah in getting up to and maintaining a high rate of speed. If the cheetah can't catch the zebra quickly, the zebra has energy reserves that will carry it well beyond the cheetah's short range.

5. Answers will vary widely.

6. Students must first identify with their animal of choice, then write a story about how they survive. Depending on the time you wish to budget, responses might range from a few paragraphs to several pages. Here are some discussion ideas to stimulate creative thinking:

a. If cats use bursts of speed to catch prey, how do dogs catch prey? (They run slow and steady until they wear the hunted animal down.)

b. Bears can run and swim much faster than humans. Large ones can reach twelve feet into a tree. If you were attacked by a grizzly bear, would you use speed, endurance, or stillness to survive?

c. When confronted by people who threaten you, how would you survive? Would you run? Would you verbally or physically return the aggression? Or would you use stillness or passivity to still their anger and violence?

BEAT IT!

Imagine you are standing in an open field. A hungry hunter is looking for you.

1. Draw what you might look like if you could escape into the trees. Explain in a few sentences how your body helps protect you.

2. Suppose you choose the rocks. Draw yourself as a different animal, and explain in a few sentences how you survive.

ROCKS

Draw and write in these spaces. Use the back if you need more room.

TREES

YOU ARE HERE

3. Design a critter that escapes over the river and into the grass. How would you get away and stay safe?

GRASSLAND

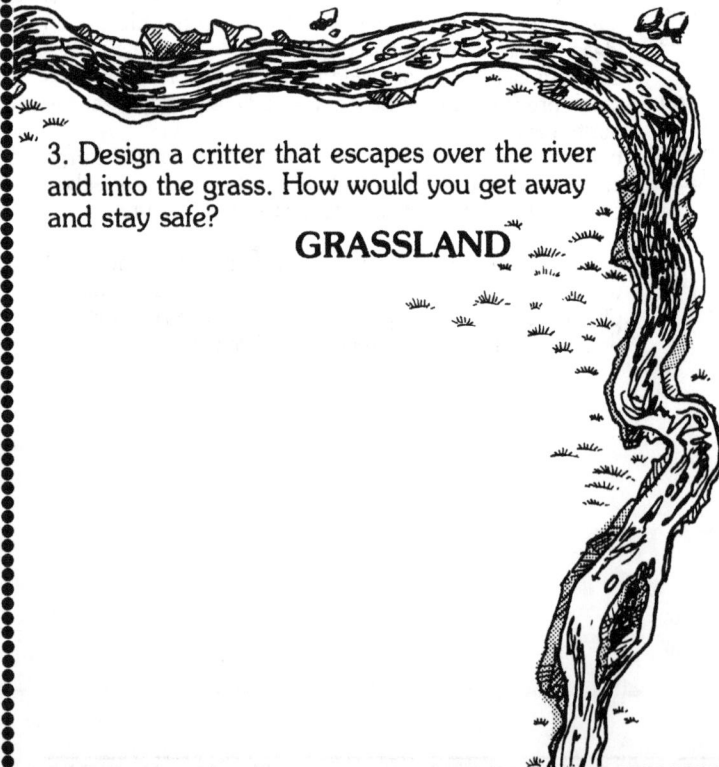

4. How might you look if you escaped into a herd? Explain how this saves you.

ANIMAL HERD

Objective

To recall many different ways that animals survive. To relate animal survival strategies to variations in habitat.

Introduction

Imagine you are standing on an ocean beach. A hungry hunter is looking for you! (Don't define the predator.) Change into an animal shape that helps you survive. What would you look like, and how do you escape?

Entertain suggestions from the class. Encourage students to describe *both* a body shape *and* the resulting benefit. Draw at least one example on your blackboard.

I scare away predators by raising my claws! If that doesn't work, I can dig into the sand to hide.

Lesson Notes

1-4. Your students are free to be any kind of animal they like, as long as they conform to natural law. They can, for example, "disappear" among the trees by looking like a tree, but they can't simply vanish into thin air.

Students should brainstorm many possible options before selecting a particular survival strategy. They might, for example, take cover in, on, under, or among the trees. Identify students who work in haste. Redirect them to add greater detail to their drawing, to make their written descriptions more complete.

Extension

Write each habitat on a separate slip of paper. Duplicate as necessary so there are as many pieces of paper as students in your class:

a windy place	a wet swampy place
a usually frozen place	a dry dusty place
an island in a small lake	a flat grassy prairie
extremely steep mountains	a deep thick forest
a steamy rain forest	the deep ocean
a city, in old buildings	near a coral reel
on the beach	near a large highway
in a swift river	

Put these slips of paper into a box or hat, then hold a drawing. Students should design an original (never seen before) creature that can survive in the environment they drew. They should present an oral report about their creation to the rest of the class.

Summarize these possible survival strategies on your blackboard:

a. FLEE: run fast, run far, jump, dodge, crawl, fly, glide, dig, swim, dive.

b. HIDE OR CONFUSE: blend in (camouflage), imitate something dangerous (mimic), play dead.

c. WARNING COLORS: sting, taste bad or poisonous, bite, smell bad.

d. PROTECTIVE COVERING: horns, armor, shells scales.

e. INTIMIDATE: make a threatening noise, look big, bad and ugly.

If materials are scarce, you might limit creative expression to pencil and paper drawings. If you have access to construction supplies — colored paper, clay, papier-mache, paint and the like — allow students to fashion more complex creatures in 3-dimensions.

Oral reports should explain in a thoughtful manner how each creature survives: how it gets food; how it defends itself. Encourage whole ideas, not one-word, dull nothings. Wild flights of fantasy are okay, as long as they are intelligently presented. Ask probing questions to expose gaps of thought. Consider trade-offs. Demand that loose ends be tied up.

After each presentation, ask the rest of the class to vote on each beast's chances for survival.

thumbs up
will survive

thumbs sideways
may survive

thumbs down
extinct

Check Point

1. I'm a monkey. My long legs help me run fast and climb high into the trees. My brown color makes me hard to see.

2. I'm a turtle. Hiding inside my shell among the rocks makes me look just like another rock!

3. I'm a snake. I'll swim across the river and sneak under the grass where no predator can see me.

4. I'm a zebra. If I run with the herd, there will be so many black and white stripes moving in so many directions, predators will get confused.

O. PLANTS

The best way to learn about plants is to grow them. Seeds are the logical place to start. Your class will soak them in water, put them in soil, then observe plant development as it happens, from germination through growth into young seedlings.

One picture, accurately drawn, can describe a plant better than 1,000 words. These activities enable students to practice important drawing skills. They make detailed to-scale drawings using a system of grids. One grid placed directly behind the plant serves as a background divider. Students then sketch the plant image, one square at a time onto a second grid. Splitting complex plant images in this manner, into simple squares, enables students to produce correctly proportioned drawings with remarkable detail. Over a period of several weeks, these scale drawings, like snapshots in time, document an accurate record of plant growth.

Each specific part of a plant has its own special name. It is not necessary to present your class with a list of words to memorize. Within the context of drawing and labeling, this plant vocabulary will develop naturally. New words become old friends, not by artificial memorization, rather, by simply practicing good science.

EVALUATION

Each question evaluates a single activity from PLANTS as numbered. Use any combination to frame a formal exam or an informal review: Copy these questions on your blackboard, construct your own ditto master, or photocopy the questions while masking out the rest of the page. Evaluate in ways that suit your own teaching style, enabling your students to learn and enjoy science.

Questions

O-1
How is a dicot seed different from a monocot seed?

O-2
An empty box has been laying in a grassy field for several weeks, open-side down. If you lifted it up and looked underneath, predict what you might see.
(a) Would the grass be green? Explain.
(b) Would it have the same average length as grass in the rest of the field? Explain.

O-3
Will seeds germinate in dry soil? Explain.

O-4
Summarize in words and pictures how each kind of seed germinates.

DICOTS?   MONOCOTS?

O-5
How are leaves from your dicot different than leaves from your monocot?

Answers

O-1
The dicot has two cotyledons that easily divide into two separate parts. The monocot has but one cotyledon and thus cannot be opened without cutting or tearing it apart.

O-2
(a) No. Without light the grass could no longer make green chlorophyll. Its color would fade to yellow.
(b) No. The grass would be longer. Rapid growth is a natural response to lack of light, increasing the possibility of growing into life-sustaining sunshine.

O-3
No. The seeds will remain dormant until moisture penetrates the seed and initiates the germinating process.

O-4
(See "Check Point" in teaching notes O-4. The development of both the pinto bean and popcorn seed are detailed in parts a-e.)

O-5
Dicot leaves have thick primary veins that branch into thinner secondary and tertiary veins. They are often lobed and intricately edged. Monocot leaves are parallel-veined, and much less varied in shape than dicot leaves.

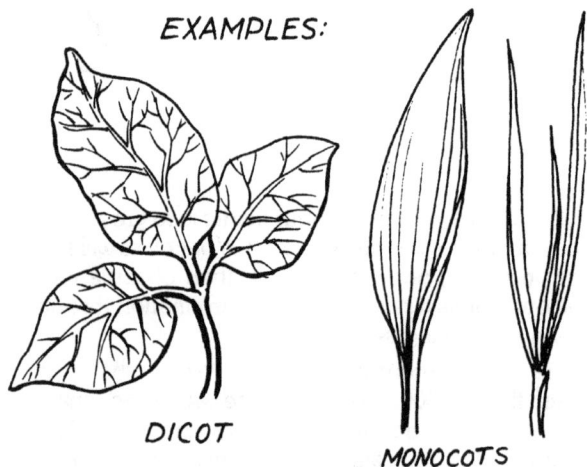

EXAMPLES:

DICOT   MONOCOTS

SEQUENCING

PLANTS may be studied at any time. It has no prerequisites.

Related Activities: **O**

MATERIALS

Here is everything your students will use for the next 5 activities on PLANTS. Materials printed in normal type are part of the core 15-things-in-a-box inventory that support all 100 activities. Materials printed in *italics* are additional local materials that you provide or ask your students to bring from home. Pencil and paper are already assumed and therefore unlisted. Each item is numbered with the activity where it is first used.

(O-1) *Pinto beans* and *popcorn*, or equivalent. See "Preparation" in Teaching Notes O-1.
(O-1) *Small jars with tight-fitting lids*, 1 per student or activity group. See "Storage of Germinating Seeds" in Teaching Notes O-1.
(O-1) Scissors.
(O-2) *Medium-sized cans.*
(O-2) *Newspaper.*
(O-2) A *water* source.
(O-2) Aluminum foil.
(O-2) Masking tape.
(O-3) Size-D dry cells.
(O-3) *Damp soil.*
(O-4) Self-adhering *plastic wrap* or equivalent. See step 3 under teaching notes O-4.
(O-5) Paper clips.
(O-5) Clear tape.

FURTHER STUDY

Use problems like these plus "extension" ideas in PLANTS to lead your students beyond worksheet activity into original research and investigation. Each discovery leads to more questions, deeper questions, better questions than these. Answering them is what good science is all about.

Think about the different ways that seeds can travel. Write a story about the adventures of some of these seeds. Collect examples to illustrate your story.

The Earth's forests are vanishing at an alarming rate. Investigate this problem and propose some solutions.

Write haiku poems that describe your feelings about the plants you have raised.

Haiku: an unrhymed Japanese poem about nature with 3 lines containing 5, 7 and 5 syllables respectively.

COMPARING SEEDS

1 CUT OUTS O-1

Put a different kind of seed on each small grid, and write its name in the box underneath.

First place seeds here...

2x ←

SEED A

2 Draw each seed, 1 square at a time, onto its larger grid.

...then draw. Label the parts.

← 2x

→ 2x

SEED B

3 Compare and contrast the *appearance* of each seed.

Size? Shape? Color? Feel?

4 Soak 15 seeds of each kind under water.

Let soak overnight.

stop!

5 Scrape off the soft (soaked) covering from 2 different seeds. Use your fingernails.

Open the seed if this is easy to do!

6 Make enlarged labeled drawings as before.

2x ←

CUT OUTS O-1

SEED A

→ 2x

SEED B

7 Compare and contrast the *structure* of each seed.

Store the seeds as directed by your teacher.

What parts make up the whole?

Objective

To compare and contrast two kinds of seeds in words and pictures. To sprout these seeds in a warm moist environment.

Preparation

These activities require two kinds of seeds. Pinto beans and popcorn seeds are recommended. If these are not available in your locality, substitute similar kinds of seeds. Test them in advance to make sure they are fertile.

One kind of seed must easily open when softened by water. Seeds of this type are called dicotyledons, "dicots" for short, because they have two cotyledons (food storage organs). Peanuts, lentils, peas, radishes, and beans all split open into 2 cotyledons.

Your second kind of seed must be a monocotyledon, "monocot" for short. Grass seeds, rice, rye, oats, wheat and corn are all difficult to split open because they have only 1 cotyledon.

Introduction

Introduce basic plant vocabulary, by sketching these seeds on your blackboard and naming the parts aloud for your students to hear. Even though these drawings specifically represent a pinto bean and popcorn, the names generalize to all dicots and monocots:

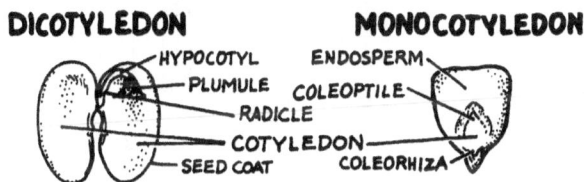

cotyledon (cot-uh-LEE-dun)
hypocotyl (HY-poe-cot-uhl)
radical (RA-duh-cul)
plumule (PLUME-yule)
coleoptile (ko-lee-AP-tul)
coleorhiza (ko-lee-uh-RYE-zuh)
endosperm (EN-duh-sperm)

Lesson Notes

1-2. These grids double the image-size of each seed. Using sharp pencils, students should have enough drawing area to capture small surface details and reproduce subtle textures.

4. Students must stop here to soak the seeds overnight. This softens some for dissection (step 5) and prepares the rest for germination. Students should proceed directly to activity O-2, then return here to finish up tomorrow.

These activities use a total of 12 seeds of each kind, 24 in all. Three extra of each variety are added (a total of 30) in case some seeds are not viable.

5-7. Dicots softened by absorbed water should open easily, preserving all internal structures intact. Monocots, by contrast, can only be divided with a knife or torn asunder. It is better to simply remove the outer covering on monocots and draw the seed as a single unit.

STORAGE OF GERMINATING SEEDS

Seeds from this activity will be used in the next 4 activities. It is important to keep them surrounded by moisture and oxygen so they will develop as rapidly as possible.

Instruct students to pour out most of the water so the seeds are only standing "knee-deep". Then seal this water inside by covering the top of each jar with a tight-fitting lid. (If you don't have lids for all your jars, seal them with pieces of plastic and rubber bands.) In tropical climates, exchange water on a daily basis to prevent mold from growing inside. Given water, oxygen and sufficient warmth, these seeds will begin to sprout, ready for use in activities O-3 and O-4.

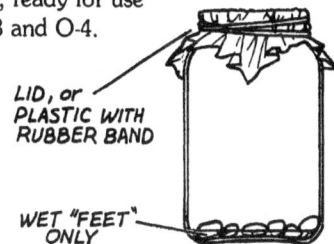

LID, or
PLASTIC WITH
RUBBER BAND

WET "FEET"
ONLY

Check Point

These drawings and descriptions are all based on pinto beans and popcorn seeds.

1-2.

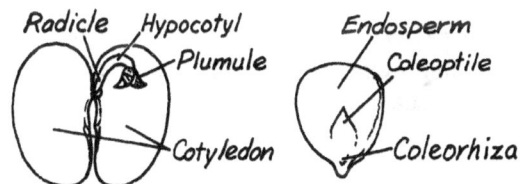

PINTO BEAN   POPCORN

3. The pinto bean has a round, smooth kidney shape. It is covered by numerous dark brown speckles and spots on a light brown background. The popcorn seed is somewhat smaller, rougher, and more angular, narrowing to a blunt point on one end. Its color varies from deep yellow to almost white.

5-6.

PINTO BEAN   POPCORN
(open)   (covering removed)

7. The bean embryo consists of two easily separated cotyledons attached to a tiny plant-like structure. Its plumule (first true leaves) and root tip are clearly visible. The part in between is called the hypocotyl. Structural details in the popcorn are not so easy to define because the single cotyledon will not easily divide. A small shoot (the coleoptile) can clearly be seen bulging through the endosperm.

NIGHT AND DAY

1 Get a can and a jar.

2 Trace around newspaper to make circles of paper.

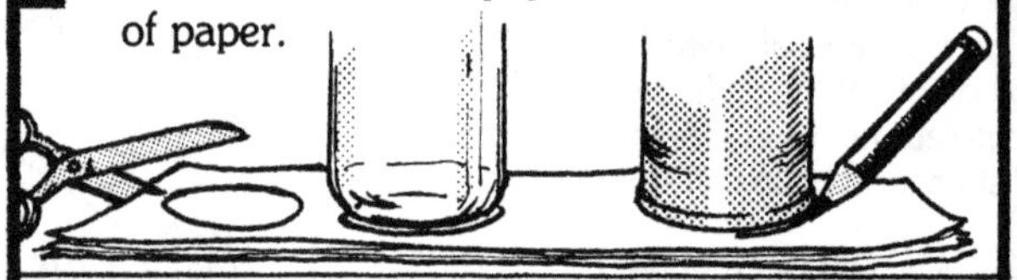

Put at least 16 circles in each container.

3 Soak the papers with water, then drain.

SOAK   DRAIN EXCESS

4 Add 3 soaked seeds of each kind to each container.

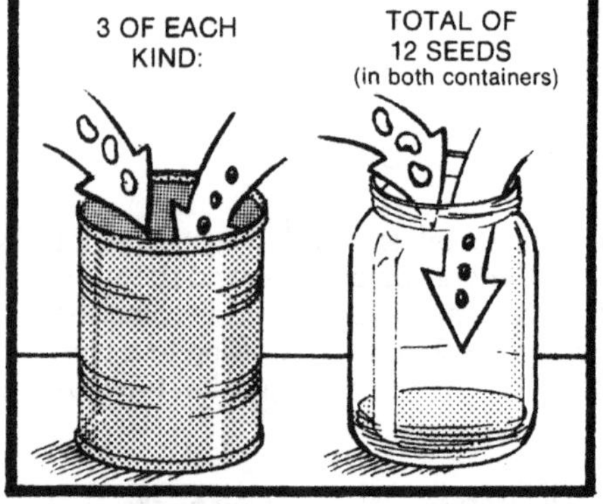

3 OF EACH KIND:

TOTAL OF 12 SEEDS (in both containers)

5 Seal airtight with foil, then tape.

AIRTIGHT!

TAPE!

6 Set aside for about one week.

stop!

7 After a week, draw exactly what you see in the can and the jar.

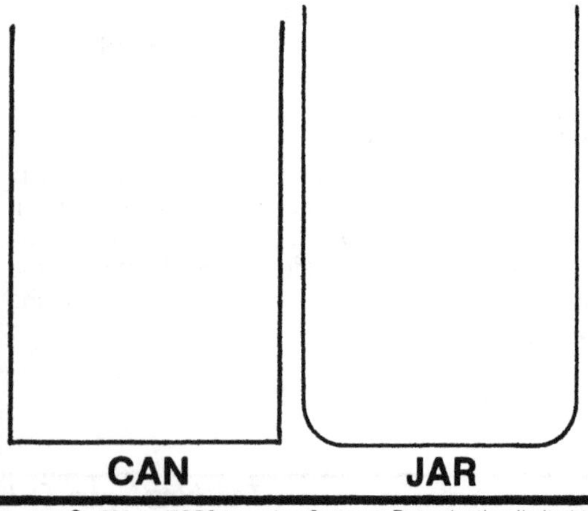

Compare and contrast light-sprouted seeds with dark-sprouted seeds.

CAN   JAR

Objective

To study the effects of light and darkness on the germination and growth of seeds.

Introduction

None required.

Lesson Notes

4. These seeds are drawn from the jar of water previously set up in activity O-1, step 4. It's OK if they have just started to soak. They can absorb more water, then germinate, in the airtight containers of step 5.

5. If there are holes in the foil, tape them as well. A tight seal is critical to insure that no water evaporates to the outside. Otherwise the small quantity of water that soaked into the newspapers (step 3) will not be sufficient to sustain the developing seedlings over the 1 week duration of this experiment.

Plastic wrap secured by rubber bands also provides a good seal. Foil should still be applied overall to block out sunlight from the can.

If you are conducting this experiment in a humid tropical climate, fast growing molds may attack the young seedlings before they have a chance to grow. If this happens, you may need to sterilize both containers before sealing the seeds inside.

6. Designate a safe out-of-the-way place in your classroom, with indirect daylight, where students can store this experiment in progress. They should continue with the remaining activities in this series, returning here one week later.

7. How will the seeds compare after seven days in a moist, *dark* environment as opposed to a moist, *light* environment? Will they both sprout? Will they look the same? If the cans have remained sealed with tape, if nobody has already looked, then who can say for sure? Everyone will be curious to find out. It's like opening a present when you don't know what is inside.

Everyone knows that plants need light to grow. It's not unexpected, therefore, to find that many students will typically predict that only the seeds in the jar will sprout. Others who think more deeply, about seeds sprouting underground in the *dark*, damp soil, may predict the opposite.

Rightness or wrongness is not the issue here. Good science is. A well-reasoned prediction that is wrong is better than a lucky guess.

Be sure to capitalize on your student's high interest by engaging them in debate *before* they open their cans and jars. When students disagree, encourage them to defend their logic with good reasons. Take a vote on who thinks this and who thinks that. Then pull off the foil.

This step only requires students to observe *how* light affects germinating seedlings. Here is a deeper explanation that begins to answer *why*:

Most seeds easily sprout in the dark. Some, but not all, actually prefer darkness. Once they have germinated, however, these seedlings must grow into life-sustaining sunlight, where they can start photosynthesizing their own food. Growing tall is thus a survival response — a race toward the light. The seedlings remain white as they shoot up, until they can tap into the sun's energy to finally assemble their vital chlorophyll molecules.

Color chlorophyll green. The leaves, cotyledons, even the stems store it in their chloroplasts. Chlorophyll enables plants to use the sun's energy to strip hydrogen from water and combine it with carbon dioxide to produce sugar. Sugar, in turn, fuels all vital functions. No light energy, no chlorophyll, no sugar, no chemical energy, no seedling, no life.

Check Point

7.

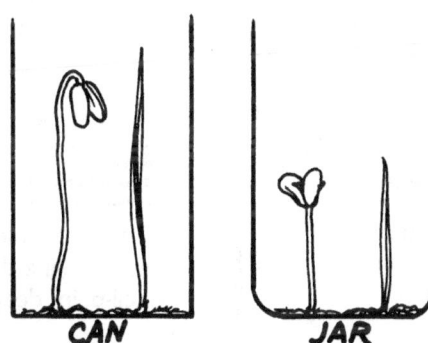

CAN   JAR

The dark-sprouted seedlings grew long and spindly, with colors that ranged from ghost white to yellow. The pinto bean's cotyledons remained in an under-developed bent-over position. The light-grown seedlings by contrast grew shorter but in more normal proportions. They ranged in color from dark green to yellow. The pinto bean's cotyledons were uplifted and opening.

FOIL MINI-PLANTERS

1 Tear off a 10 cm strip of foil. . .

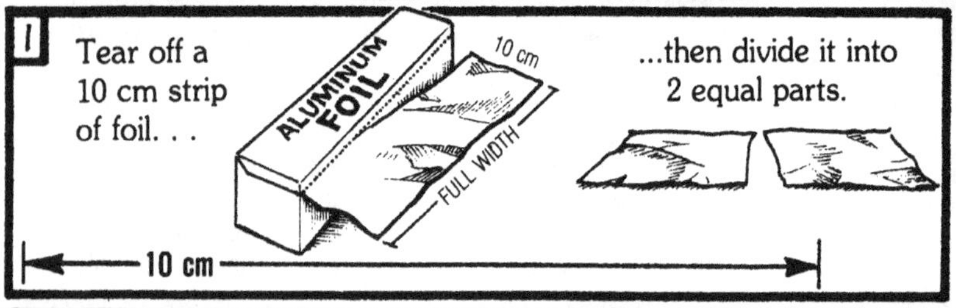

ALUMINUM FOIL
10 cm
FULL WIDTH

|← 10 cm →|

...then divide it into 2 equal parts.

2 Fold each piece in half the long way.

3 Stick masking tape around the positive end of a dry cell, even with the edge.

POSITIVE END

TAPE AT EDGE

4 Wrap your foil around this dry cell so the folded side just meets the tape. . .

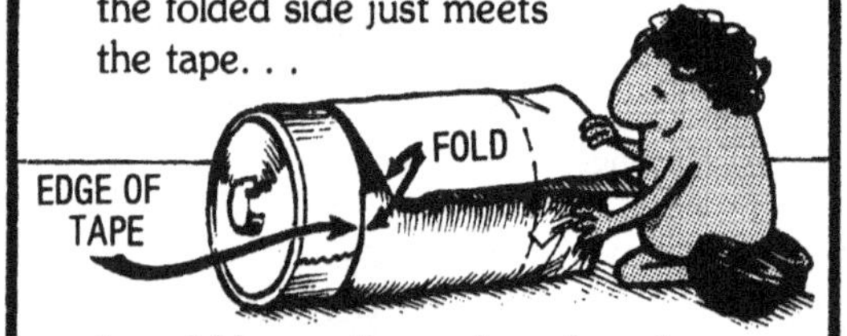

EDGE OF TAPE

FOLD

...then fold over the ends and push them flat against your table.

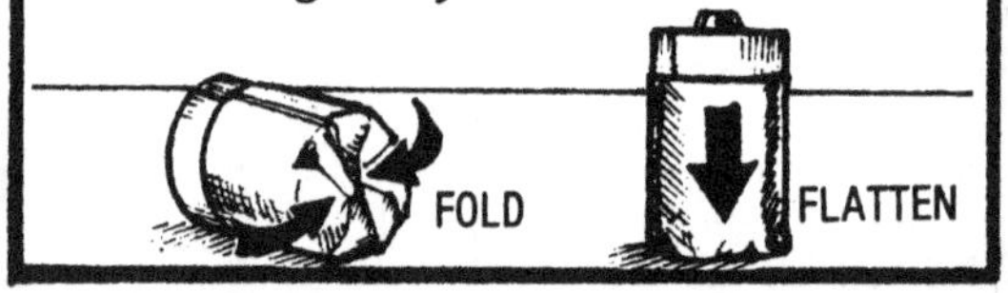

FOLD   FLATTEN

5 Fasten the top edge of the seam with a small piece of masking tape.

Pull out the dry cell, then make another planter with the other foil.

6 Fill each planter brim full with moist soil.

DAMP SOIL

7 Poke pencil holes into each planter. Make them pencil-point deep.

PENCIL-POINT DEEP

8 Use sprouted seeds from before. Drop 2 of the same kind into each hole.

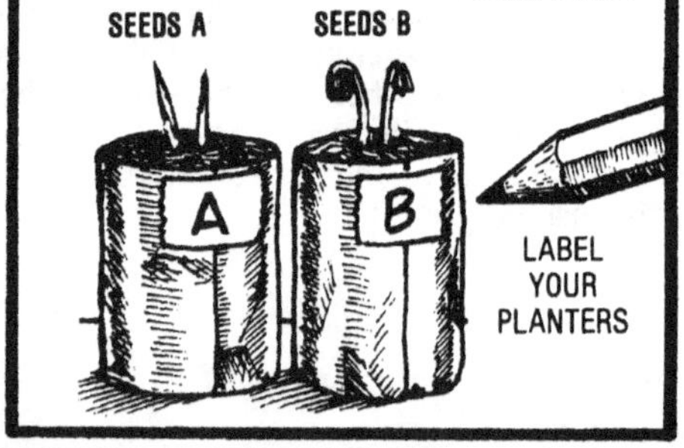

SEEDS A   SEEDS B

A   B

LABEL YOUR PLANTERS

9 Cover with soil, then stand both planters in a lid.

A   B

Add some water.

10 Always keep water in the bottom of the lid.

Don't let the lid go dry!

A   B

Objective

To construct an inexpensive, low-maintenance, space-conserving system for growing plants.

Introduction

Prepare a set of mini-planters in advance and put them in your own watering lid. This will familiarize you with the directions and provide a model for your students to follow. Moreover, if the seeds in any planter should fail to germinate, you'll have these extras to take their place.

Lesson Notes

1. Younger students may not be able to evenly tear pieces off the roll of aluminum foil. You can speed up this process by pre-tearing these sheets yourself to 10 cm widths.

3-4. This masking tape defines the ideal height of the planter. There will be just enough foil overlapping the end of the dry cell to form a solid bottom.

5. This tape serves as a label in step 8, identifying which seeds were deposited in each planter.

6. Notice that students fill each container with soil that is already moist. Premix perhaps a quart of soil plus water in a large bowl. Add enough water to make the soil damp, but not sloppy wet.

7-8. Choose seeds from activity O-1 that have already sprouted, if available. Unsprouted seeds, even dry ones, may also be substituted. But they will require a longer time to germinate and grow above the soil.

Punch holes in the soil to their proper pencil-point depth, then place the germinating seeds so their emerging radicals point down.

10. Add water to the *lid*, not the planter. It will seep through the folds of aluminum at the bottom of the planter and soak up into the soil. This stimulates the roots to grow down into the moisture, thereby anchoring the young seedlings firmly in the soil. Because the lid is shallow, overwatering is impossible and proper drainage is assured.

How long the seedlings can survive between waterings depends, to a great extent, on humidity. The lid might possibly dry out for a day or two with no adverse effect, as long as the soil inside retains sufficient moisture.

Check Point

None required.

NAME: _____ CLASS: _____

SPROUTING SEEDS

1 Trace the outline of a jar lid on at least 8 layers of newspaper...

Cut out the circles.

2 Put the paper circles into the lid...

Soak them with water and drain.

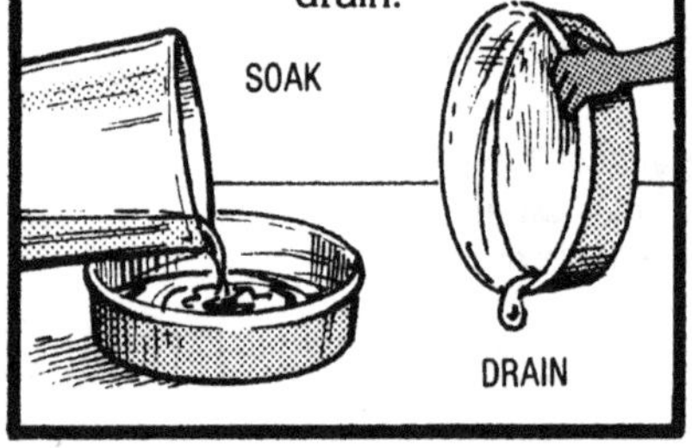

SOAK

DRAIN

3 Put 3 sprouted seeds of each kind on the wet paper...

SPROUTED SEEDS

Seal the lid airtight.

This keeps the seeds warm...

4 Look at your sprouting seeds each new day. Record, in words and pictures, how they grow.

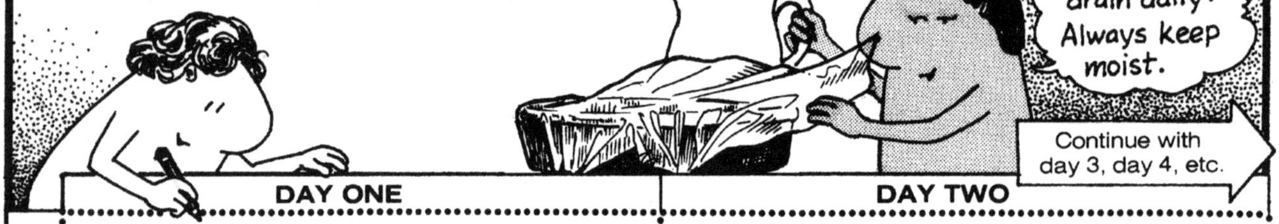

Rinse and drain daily. Always keep moist.

Continue with day 3, day 4, etc.

	DAY ONE	DAY TWO
SEEDS A		
SEEDS B		

Objective

To record in words and pictures the daily development and growth of germinating seeds.

Introduction

None required.

Lesson Notes

1. Use the lid from the seed germination jar, if available, previously set up in activity O-1.

3. Select 3 sprouts of each kind that have the most advanced development. Discard remaining seedlings or plant them outside.

The purpose of this step is to provide a moist, warm growing environment for seedling development that is accessible for easy viewing and drawing. A lid is suitable *if* it has enough depth to accommodate the growing plants and can be sealed air tight after daily observations. It might be wrapped in self-clinging plastic, covered in plastic and rubber banded, or even placed inside a plastic bag. The covering must be somewhat transparent to allow in enough light for photosynthesis. Opaque aluminum foil is not a good substitute.

Other systems that work are wide shallow jars with lids and empty styrofoam egg cartons. Cut the lids off the egg cartons, line them with a paper towels and cover with plastic wrap. Recycle whatever is easily available and make it work.

4. Students don't need to draw all 3 seedlings of each kind unless they want too. They can concentrate on drawing only one of each kind — the best developed and fastest growing. Extras are provided to insure that at least some survive.

Encourage students to take all the time they need to make drawings that accurately reflect true proportions and capture as much detail as possible. One-half hour per day is not too long to spend sketching the daily development of two seedlings. A sharp pencil and good eraser are essential to hone and refine each drawing. If you make accurate drawings a big issue, students will reward you and themselves with excellent work.

Track seedling development for at least a week or even longer, as long as interest remains high.

Check Point

Beans and popcorn develop in the sequence illustrated. Individual growth rates, of course, depend on the particular seeds you use and on general growing conditions, especially temperature.

DEVELOPMENT OF A PINTO BEAN

a. primary root pokes out

b. root hairs develop

c. secondary roots branch out

d. hypocotyl lengthens

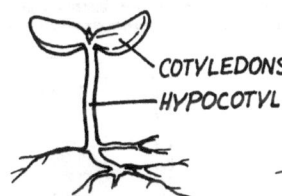

e. cotyledons unhook and open

f. first true leaves unfold

DEVELOPMENT OF A POPCORN

a. primary root pokes through coleorhiza

b. root hairs develop

c. coleoptile pushes up

d. rapid extension of primary root

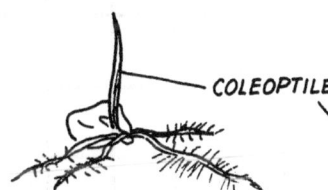

e. secondary roots develop

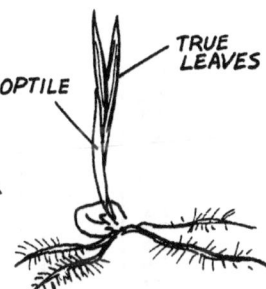

f. first true leaves emerge from coleoptile

GRID WORK

1 Cut out the grid and write your name in the box. Fold back both sides along the dashed lines.

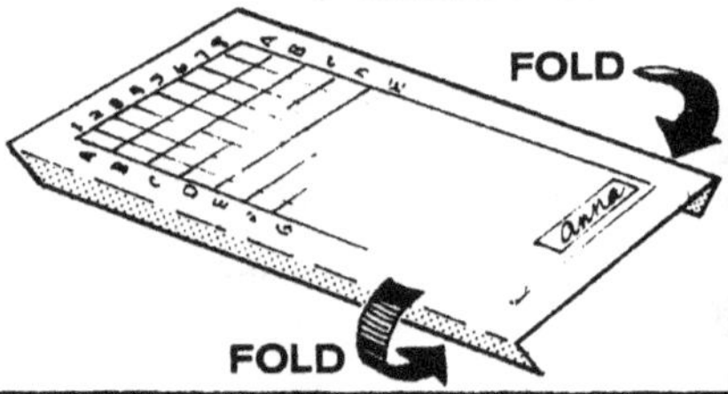

FOLD

FOLD

2 Bend a paper clip to a right angle.

90°

3 Tape this paper clip behind the grid to make it stand up the long way.

STANDS ALMOST STRAIGHT!

4 Set each planter from O-3 in front of this grid. Draw what you see (square for square) on grid paper.

CUTOUTS O-5

LABEL your drawings.

Make **WEEKLY** drawings.

Draw the **SAME VIEW** each week.

CUTOUTS O-5

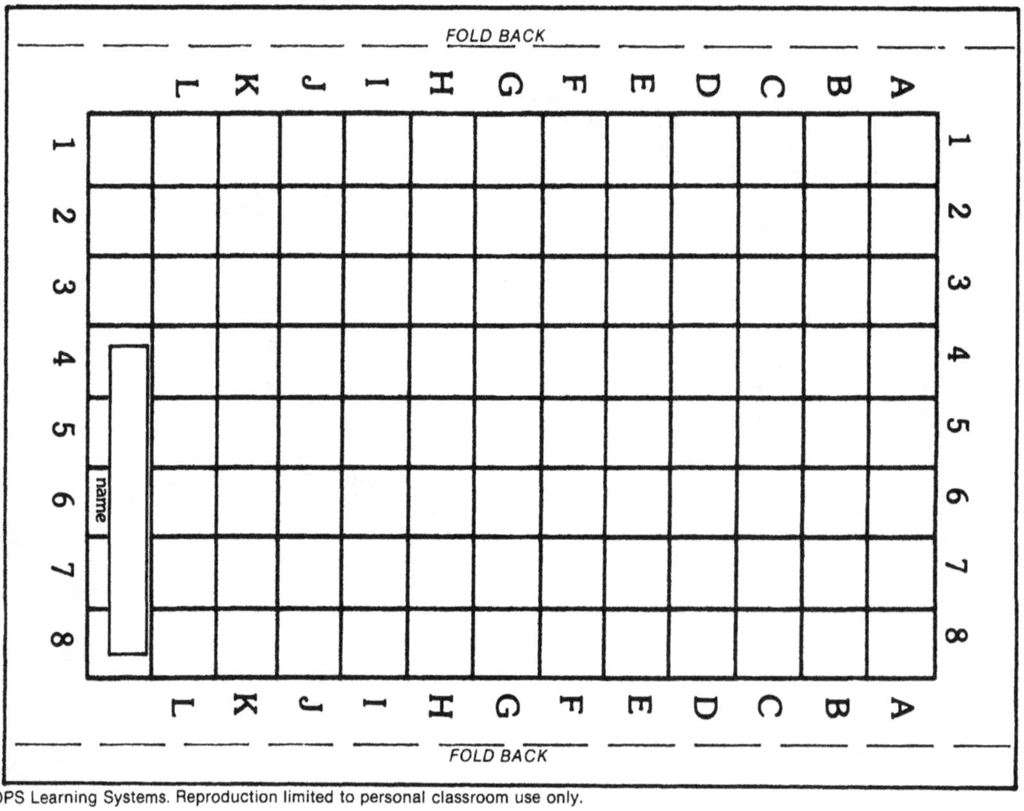

DRAWING GRID

FOLD BACK

L K J I H G F E D C B A

1 2 3 4 5 6 7 8

name

L K J I H G F E D C B A

FOLD BACK

Objective

To accurately track the growth of seedlings over 3 weekly intervals on a drawing grid.

Introduction

None required.

Lesson Notes

4. Place this vertical grid directly behind each foil planter from activity O-3, so the leaves are presented in full view. Then draw what you see on the supplementary grid sheet. This 6-grid supplement is available as a consumable cut-out or as a reproducible line master in the back of this book.

To minimize distortion, it's important to maintain a heads-on perspective, viewing each plant directly in front of the background grid. Each week, as students complete a new grid with a new drawing, remind them to turn their planters to the *same* orientation they used before. This makes it easier to compare weekly growth and development, from one picture to the next.

The task is to draw one square at a time, transferring the small part of the plant you see in each background square to the corresponding (and slightly reduced) grid. (Letters and numbers at the top and sides link squares in both grids.) By dividing complex plant images into more manageable pieces, students can capture overall proportions with greater accuracy.

To reinforce vocabulary, insist that students label all drawings.

Extension

As students continue to observe their plants, certain tropisms (growing responses) will become apparent. The leaves and stems of plants exhibit a "positive phototropism" because they tend to grow toward sun light. They also have a "negative geotropism" because they grow away from the Earth.

The vocabulary of tropisms is easy to understand, once students know the root words. To start, write these simple definitions on your blackboard:

tropism:
turning response
positive:
toward
negative:
away from
photo:
light
geo:
Earth
hydro:
water

Now, show how these words hook together into a vocabulary of plant movement. Write each composite word on your blackboard, asking volunteers to guess the meanings:

positive hydrotropism:
turning toward water

positive phototropism:
turning toward the light

positive geotropism:
turning toward the Earth

negative geotropism:
turning away from the Earth

Ask which parts of a plant might exhibit these responses:

positive hydrotropism:
roots

positive geotropism:
roots

negative geotropism:
hypocotyl, stems

positive phototropism:
leaves, stems, hypocotyl

To make tropisms a natural part of each student's vocabulary, try this peppy drill: Designate a faucet or bucket within your classroom to represent water; the windows as a symbol of light; the floor as the Earth. Then define tropisms by pointing your hand. If you say "positive phototropism," for example, point to the window. If you say "negative geotropism," point to the ceiling (away from the floor). Once your students catch on, present only the tropism (say it aloud, or use flashcards to speed up the tempo), while your class responds with the correct hand motion.

Finally, reverse the procedure. You point to a particular area in your room, while your class writes the description of that particular tropism on a piece of scratch paper. If you point away from the faucet, for example, your students would write "negative hydrotropism." Continue until your class becomes reasonably fluent in the vocabulary of plant tropisms.

Check Point

Maturing pinto beans and popcorn plants look similar to these:

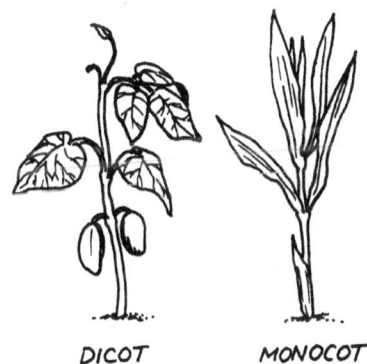

DICOT   MONOCOT

METRIC SQUARES

CUT ALL DASHED LINES

VOLUME		MASS		LENGTH	
ONE gram	As heavy as 1/2 bottle cap	.001 kg	ONE kilogram	As far as a 10 minute walk	.001 m
As heavy as a bread crumb	.001 g	ONE centimeter	As heavy as 2 paper clips	10 mm	ONE millimeter
1000 m	ONE liter	As wide as a doorway	.1 cm	ONE centimeter	As long as your leg
ONE milligram	As long as a fingernail	1000 mm	ONE gram	As much as 4 glasses of water	.001 km
As wide as a pinhead	100 cm	ONE meter	As much as 15 drops of water	.001 𝑙	ONE meter
.01 m	ONE kilometer	As thin as a coin	1000 g	ONE millimeter	As long as a monkey's tail
ONE meter	As thick as blackboard chalk	1000 mg	ONE milliliter	As heavy as a chicken	1000 ml

Copyright © 1988 by TOPS Learning Systems. Reproduction limited to personal classroom use only.

CUTOUTS

F-6

LENGTH

km

m

cm

mm

MASS

kg

g

mg

VOLUME

l

ml

valley

peak

valley

peak

METRIC CARD HOLDER

CUTOUTS I-4 ELECTRO ⚡ SQUARES

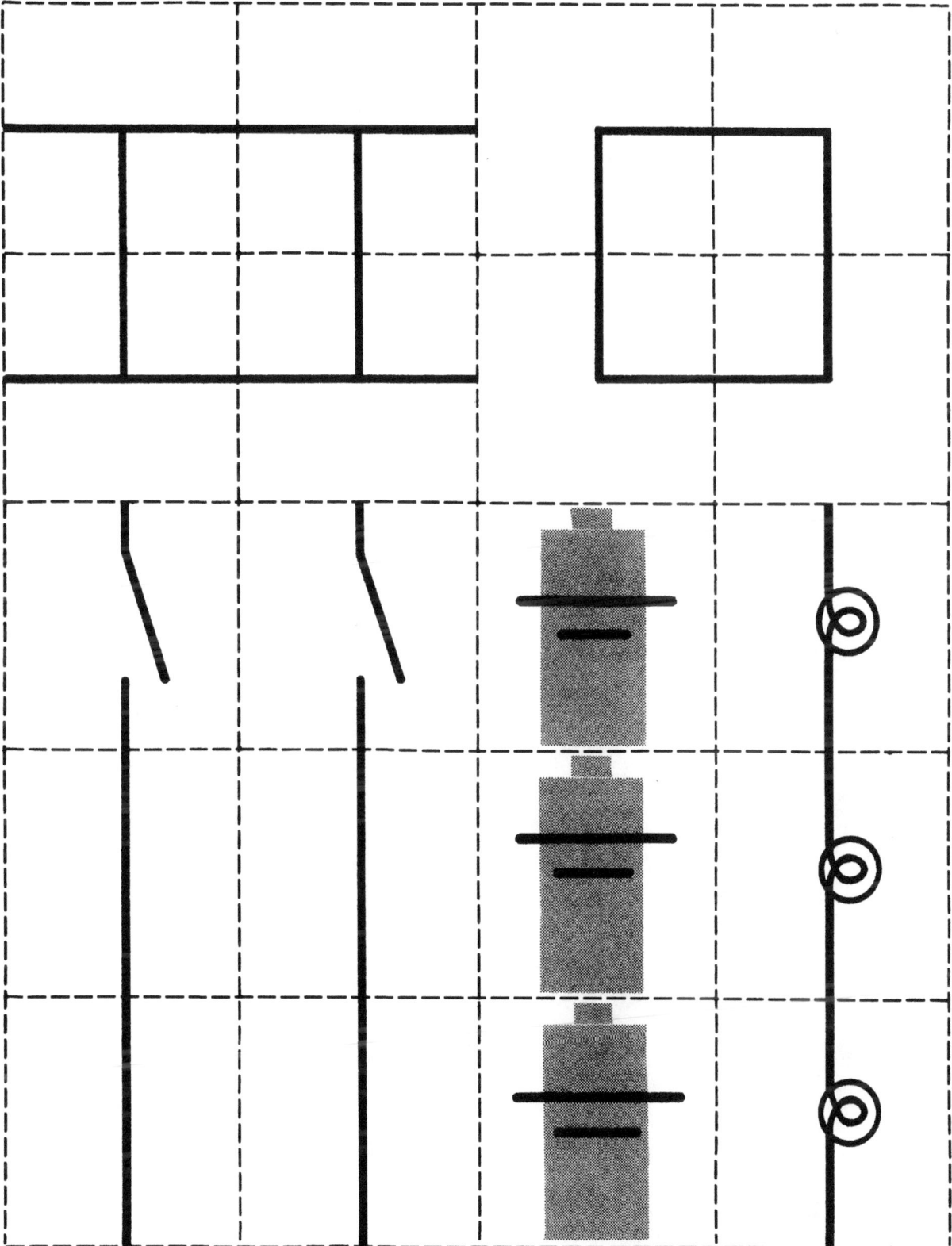

ADAPT-A-BIRD BODY PARTS

CUTOUTS N-3

BODIES:

NECKS:

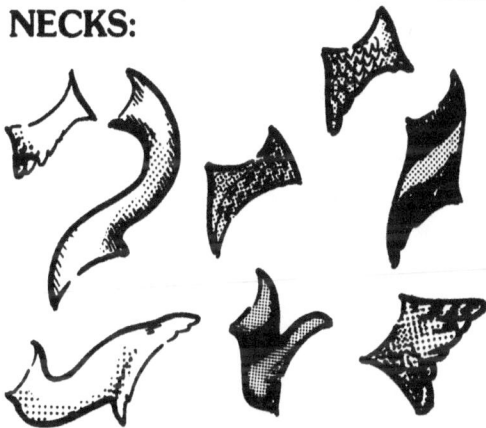

HEADS & BEAKS:

LEGS: (Don't cut apart)

Choose 1 body part from each box to make each bird.

TAILS:

FEET:

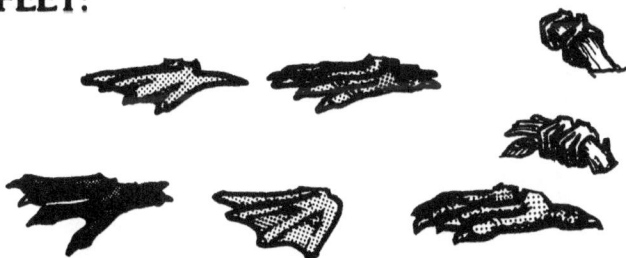

Development after...

...1 week

...2 weeks

SEEDS A

SEEDS B

www.ingramcontent.com/pod-product-compliance
Lightning Source LLC
Chambersburg PA
CBHW081057220326
41598CB00038B/7132